IMAGES
of America

NEW YORK FIRE PATROL

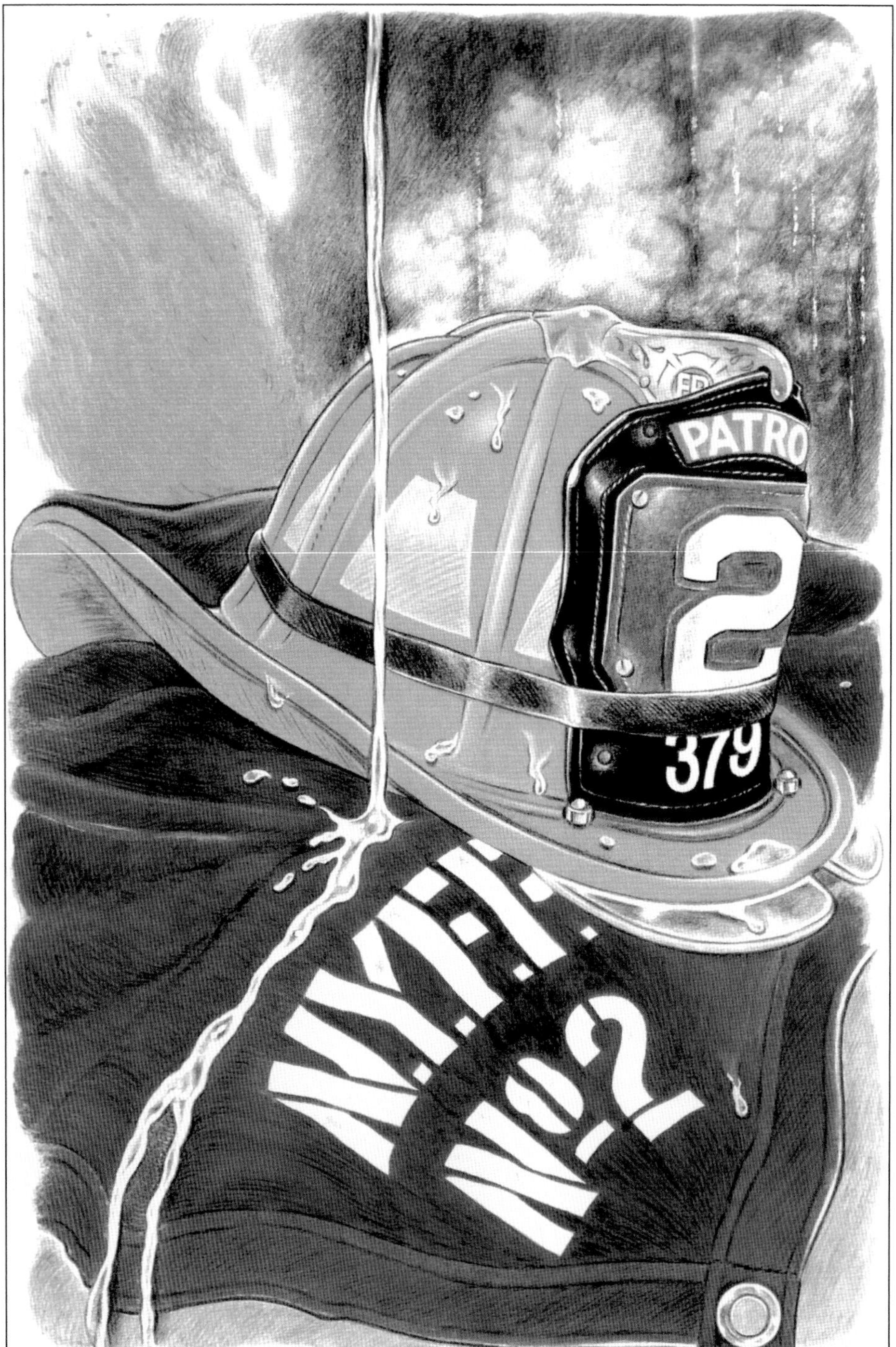

Tools of the Trade. For the past 170 years, the elements of the fire patrol have remained unchanged. Fire, smoke, and water are the enemy of the men who wear the red helmets and carry the green salvage covers. Since 1839, they have battled to save life and property. This is an original painting by patrol enthusiast Simon Gispert. (SG.)

IMAGES
of America

NEW YORK FIRE PATROL

Timothy E. Regan

ARCADIA
PUBLISHING

ISBN 978-0-7385-3874-7

Published by Arcadia Publishing
Charleston, South Carolina

Printed in the United States of America

Library of Congress Catalog Card Number: 2005926948

For all general information contact Arcadia Publishing at:
Telephone 843-853-2070
Fax 843-853-0044
E-mail sales@arcadiapublishing.com
For customer service and orders:
Toll-Free 1-888-313-2665

Visit us on the Internet at www.arcadiapublishing.com

Dedicated in memory of patrolman Keith M. Roma who was killed in the performance of his duties at Manhattan Box 5-5-8087 on September 11, 2001, and to the following members killed in the line of duty since 1853: J. S. Carman, 1853; H. M. Luke, 1858; S. Morgan, 1859; T. Mangan, 1862; J. H. Sarbach, 1863; P. Moran, 1864; J. T. Stevens, 1878; J. Spellman, 1883; C. H. Smith, 1884; Lt. J. Murray, 1888; J. Spellman Jr., 1891; A. Milner, 1893; Sgt. G. W. Dennis, 1906; J. A. Tierney, 1906; J. M. Fallon, 1908; H. Burgess, 1910; G. Ganz, 1912; Capt. P. J. Cornwell, 1916; J. J. Carroll, 1922; F. G. Wohm, 1924; E. Brennan, 1941; H. Stanton, 1942; F. C. Lehman, 1949; D. P. Shea, 1949; W. Bohlman, 1952; M. McGee, 1958; J. F. Devine, 1958; L. Brusati, 1958; M. R. Tracy, 1958; and D. Quirk, 1964.

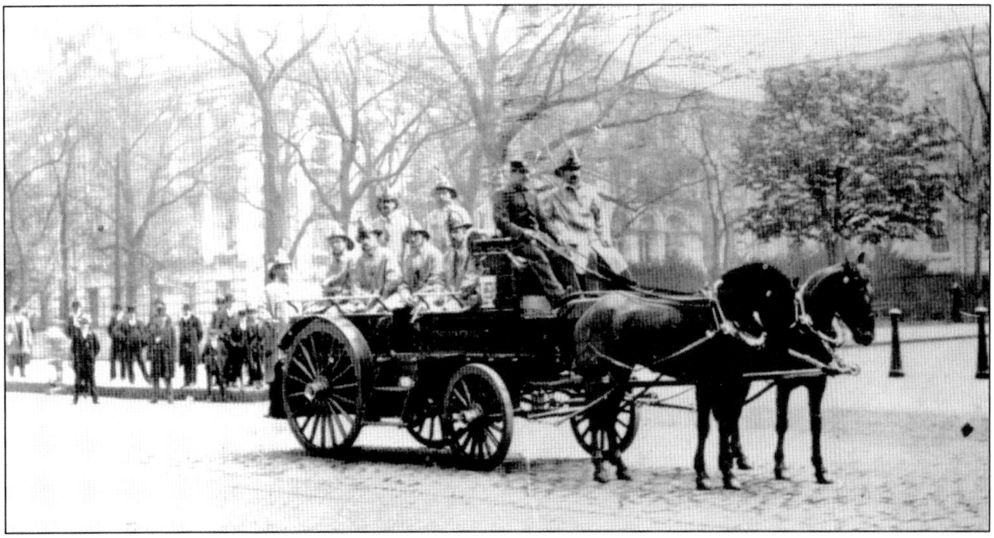

THE PATROL, 1901. The New York Fire Patrol (NYFP) is depicted here during a parade in the early part of the 20th century. (JK.)

Contents

ACKNOWLEDGMENTS

The author wishes to gratefully acknowledge the following people. Joe Krasko, whose documents and photographs formed the basis of this book. Since his joining the patrol in 1974, Krasko has served as the unofficial historian of the New York Fire Patrol and has an outstanding knowledge of this organization. Larry Wagonfeld who offered his extensive photograph collection. Paul Wormsley, former patrolman and renowned fire apparatus modeler, for his generous donation of photographs. Frank Lennon for his pictures and stories amassed over his 34 years of service. To Jack Lerch, honorary battalion chief of the FDNY for allowing access to the fire patrol files at the Mand Library; Larry McGuiness for his rare photographs and his detailed knowledge on fire apparatus; Paul Curran for his photographs, stories and encouragement; and Dennis Kenny for his collection which first sparked the idea of doing this book. Thanks also to Simon Gispert, international fire patrol historian, who provided his excellent graphics from his book *Salvage Corps and Fire Patrols*; Jim Tolfree for his images; and Tom McGee, for his pictures and anecdotes. Frank Walters is thanked for his historic data, Joe Crawley, Brian Mulvey, Ed Moran, Capt. Kevin Keegan, and Mike Krempasky. Thanks to my partner in crime, Michael Angelini (future captain, Fire Patrol No. 2). Also thanked are Pat Mooney, Carl Oreggio, Ray Weintraub, Chris DeVivo, Eric Schultz, John Ryan, Mike Melendez, Vic Melendez, John Sottille, and of course John McGee. Thanks also go to Past Chief Ray Carroll and Chief Rich Heffernan, former New York State Insurance Commissioner Gregory Sirieo, vice president Joseph N. Fiorello Sr. of the New York Board of Fire Underwriters, and the Board of Fire Underwriters members. From the beginning my personal public relations agent, researcher, and marketing advocate has been John Sheehan, lieutenant of patrol No. 3. John, I couldn't have done it without you! Lastly, to my wife Patti, who once again watched patiently, as I worked on yet "another project." Thanks Patti.

INTRODUCTION

On the pages that follow is the story of the New York Fire Patrol, a story in pictures and words that traces the very last paid Fire Salvage Department in the United States, and proudly the oldest paid fire service in continuous operation in America. Since 1839, the New York Fire Patrol (NYFP) has endeavored to save property from the destructive forces of fire as well as the ruinous effects of the water used to extinguish blazes. The NYFP today consists of three fully operational salvage companies operating from two firehouses in Manhattan and one in Brooklyn. With a force of just over 100 men, the patrol operates 24 hours a day every day. Although a completely privately operated department, its service is exclusively provided to the public, as chartered by New York state law. Of course services are provided whether property is insured or not. The patrol is an organization built on history and pride and through these pages a glimpse into that past can readily be seen.

The story of the New York Fire Patrol is one that can trace its genealogy to before the American Revolution, when the Hand-in-Hand Fire Company was organized for the sole purpose of protecting property and conducting salvage operations in the old city of Manhattan. This company was soon followed by other salvage companies including the Heart-in-Hand in 1781, the Friendly Union Fire Company in 1800 and finally in 1803 the Mutual Assistance Bag Company was formed. These four companies formed the basis of salvage operations in New York, and by 1819, the fire insurance industry had joined forces when eight companies formed the Salamander Society to coordinate the role of fire insurance in the city. In 1826, the Salamander Society was dissolved and its members formed the Association of Fire Insurance Companies, and in 1835, the association had hired four men to act as the "fire police" whose duty was to "attend all fires and protect the interests of the Fire Underwriters by preserving property exposed to fire and damage thereto by water." Each man was paid $250 annually, at which time they became the first paid firefighters in the City of New York. However, the ability of four men to cover the city's mercantile district each and every night was not an easy one, and on December 16, 1835, the city experienced a conflagration like no other. At 9:00 in the evening, a night watchman, most likely employed by the association, turned in an alarm on Merchant Street. Fifteen hours later, over 500 buildings lie in smoking ruin; the heart of the city destroyed by fire. From 1836 to 1838, additional men were hired and the organization was then known as the Insurance Patrol.

It was not until 1839 that the association realized that a larger force of men would be needed for salvage work, and in November of that year, 40 men were added to the roles. Nearly all of the men were colorful members or ex-members of the city's Volunteer Fire Department, as it was essential for the fire policemen to have a background in firefighting. In the first year of operation, $17,917 was allotted for expenses. Each man was equipped with a bright red leather fire cap, whistles, and rattles. Up until this time all salvage companies used sturdy linen bags to gather personal and commercial property during times of fire. The salvage bags were kept by the members, and each of them were adorned with the owner's initials, similar to fire buckets of the time. The first year also saw the first apparatus purchased for $53, which was kept inconveniently on the second floor of a building on Dutch Street. In 1845, the first salvage covers were purchased, and this reversed the operations of the patrol from removing goods from endangered buildings to covering exposed property in buildings with large waterproof, oil canvas covers. Around 1850, the Association of Fire Insurance Companies was reconfigured into the Board of Fire Insurance Companies, who then took over the financing and supervision of the patrol. In their 1853 minutes it states, "The Patrol shall aid the fire department in preserving the exposed property by covering the same with oiled cloths, by removal or otherwise so as most effectively protect the same from damage." At about this time, the term patrolman was coined, replacing the title of watchmen. In 1867, the fire patrol was granted legislative authority through its charter with the State of New York which empowers the patrol "to provide a patrol of men and a competent person to act as Superintendent, to discover and prevent fires, with suitable apparatus to save and preserve property and life at and after a fire. . . . " At this time, the present Board of Fire Underwriters was established for the control and management of the patrol. Up until this time, insurance companies paid into the patrol fund on a voluntary (and often not at all) basis, the new law also levied a mandatory per centum tax on all fire insurance premiums written in the city, ensuring the permanent funding of the fire patrol.

Over the years, additional units were added until the New York Fire Patrol (NYFP) eventually numbered 10 companies with a force of nearly 400 men. There is no way to tell the story of every man who gave his life for the protection of property in the great City of New York, or to tell the individual tragedies of the men who gave their lives in the line of duty. The New York Fire Patrol is an incredible organization which has withstood the test of time and has evolved over its three centuries of service. It has outlived all of the other Fire Insurance Salvage Departments in the United States, since Philadelphia closed their doors in 1962. However, for every member of the patrol there is a sense of pride knowing that saving personal property from destruction is a noble one and that occasionally the patrol saves lives in the performance of their duties. May the fire patrol continue to serve the city of New York for many more years just as it has proudly done for nearly two centuries.

One

THE JOB

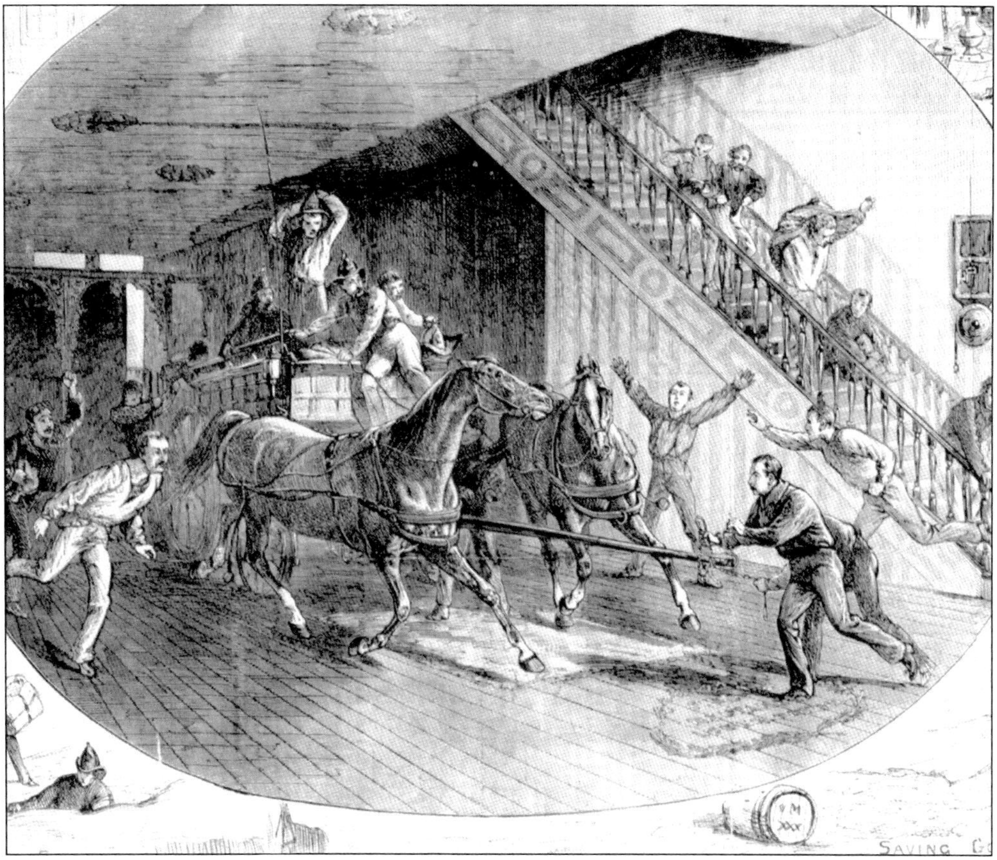

FIRE PATROL 1, MURRAY STREET. This 1876 wood engraving appeared in *Harpers Weekly* newspaper. It shows the men turning out quickly for the alarm, removing goods from burning buildings while safeguarding them, and covering merchandise from fire, water, and smoke damage. (LW.)

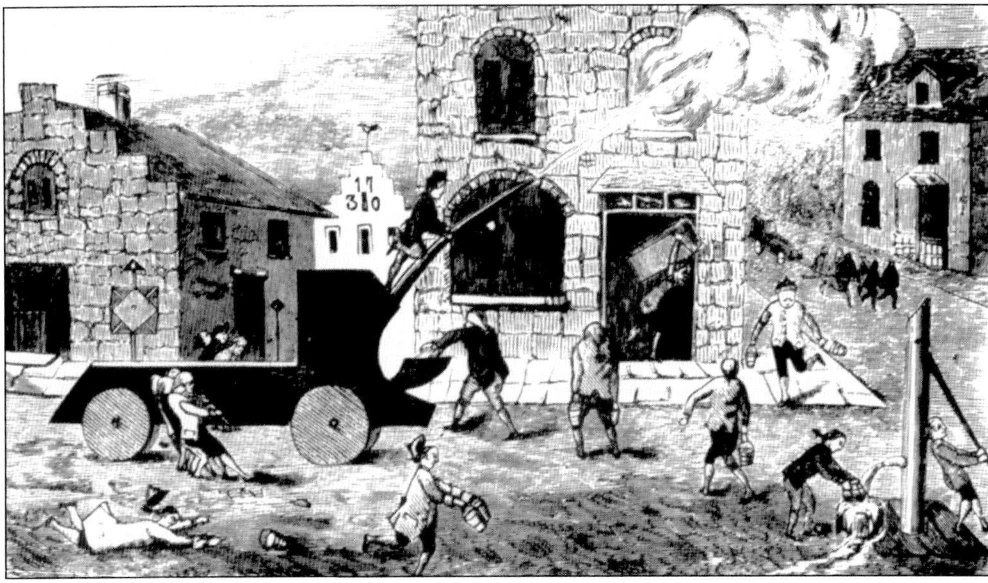

A FIRE IN NEW AMSTERDAM, 1730. This early wood engraving depicts a valiant group of volunteers battling a fire in an old Dutch dwelling. This image is notable as it depicts a lone salvage man removing a trunk from the burning building. It is the oldest known view of salvage operations in New York City. (NYFP.)

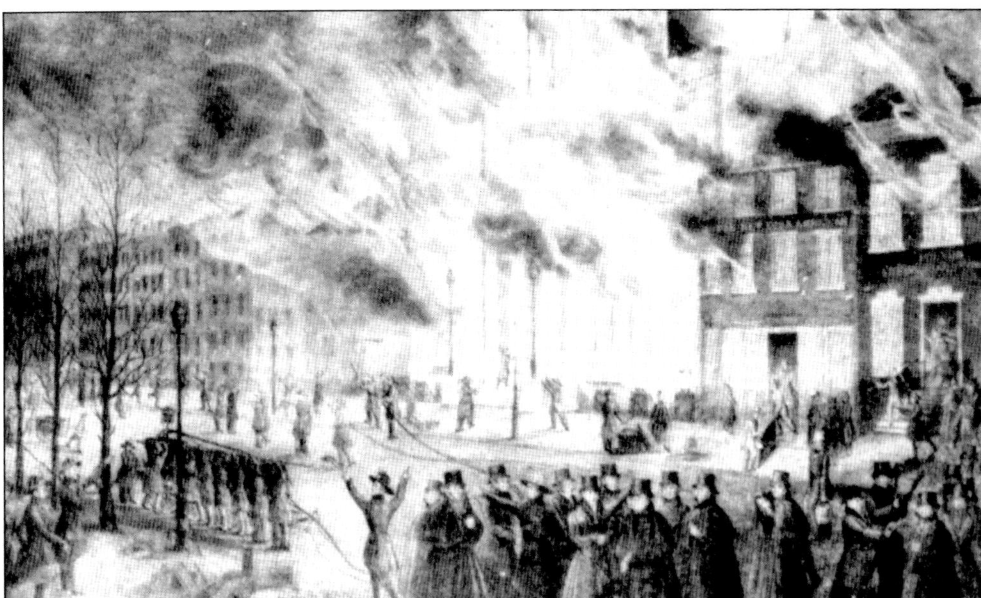

GREAT FIRE OF 1835. As flames consume the Merchants Exchange building, the early fire police, merchants, and volunteers are acting fast to remove goods from endangered buildings. The fire of 1835 led to an expansion of the fire police and within four years the Fire Insurance Patrol had been officially organized.

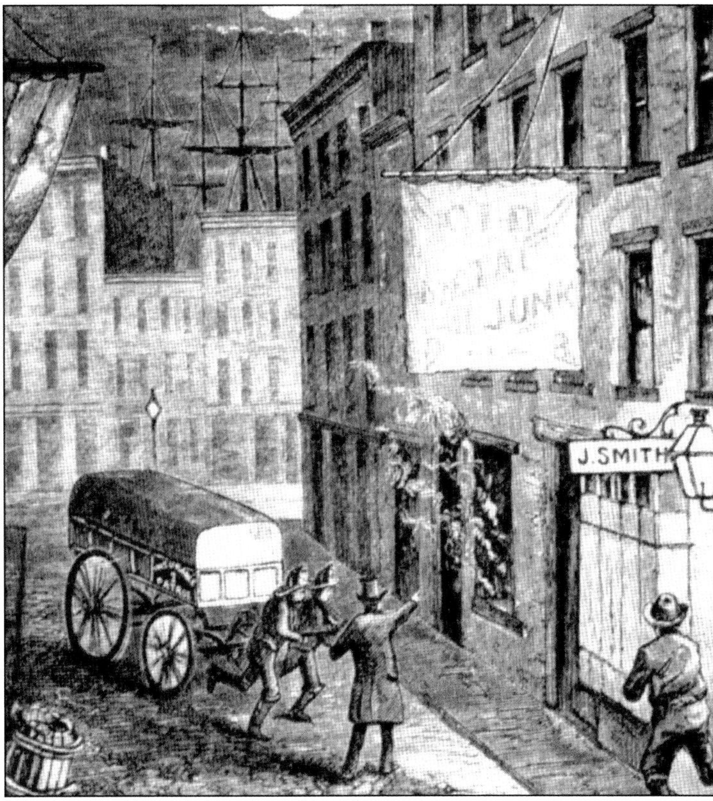

THE INSURANCE PATROL, C. 1840. The earliest known view of the New York Fire Patrol (NYFP) in action is this wood engraving from 1880. It shows the small wagon of the patrol arriving first at a commercial building fire near Peck Slip in lower Manhattan. At this time, the patrol was on duty only at night. The watchmen were to patrol the streets for fires and if they found one, they had to return to the Dutch Street quarters, and retrieve the wagon, then proceed back to the fire. (NYFP.)

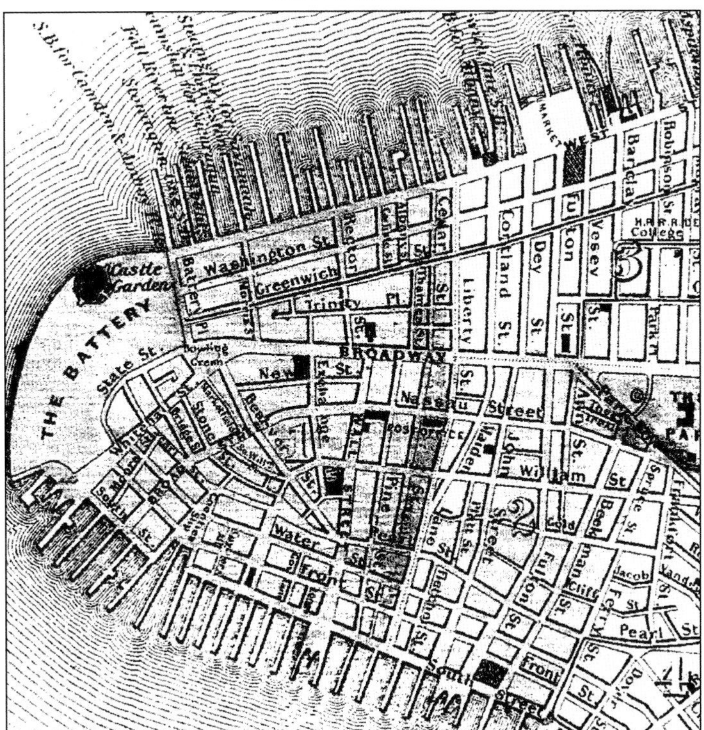

PATROL DISTRICT, 19TH CENTURY. This 1855 map of lower Manhattan encompasses much of the patrol's response district before 1867. During the first half of the 19th century, the principal business center of the city was concentrated below Canal Street. The patrol was kept busy with the many fires in warehouses, packing houses, even aboard ships which lie on the many piers surrounding old Manhattan. In 1855, the central patrol was organized (Patrol 2) and was responsible for the area up to 23rd Street. (TER.)

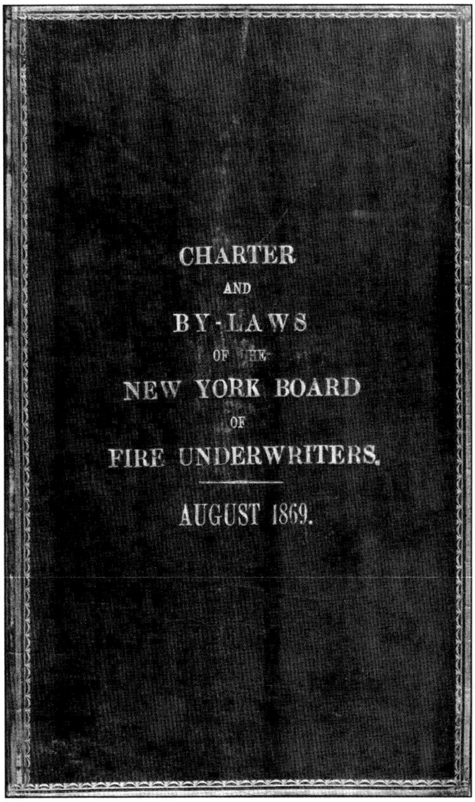
CHARTER
AND
BY-LAWS
OF THE
NEW YORK BOARD
OF
FIRE UNDERWRITERS.
AUGUST 1869.

FIRE PATROL CHARTER, 1869. The fire patrol had been operating for many years with the understanding that they could enter premises on fire, remove goods, protect the same, and owners would be grateful of the service. During the 1850s, property owners began questioning the patrol's authority to enter private property. This caused a dilemma, for if the patrol could not enter private property during times of fires, what good was their existence? These incidents led directly to the New York State legislature creating the charter for the operation of a private fire patrol to do public service in the city of New York. (TER.)

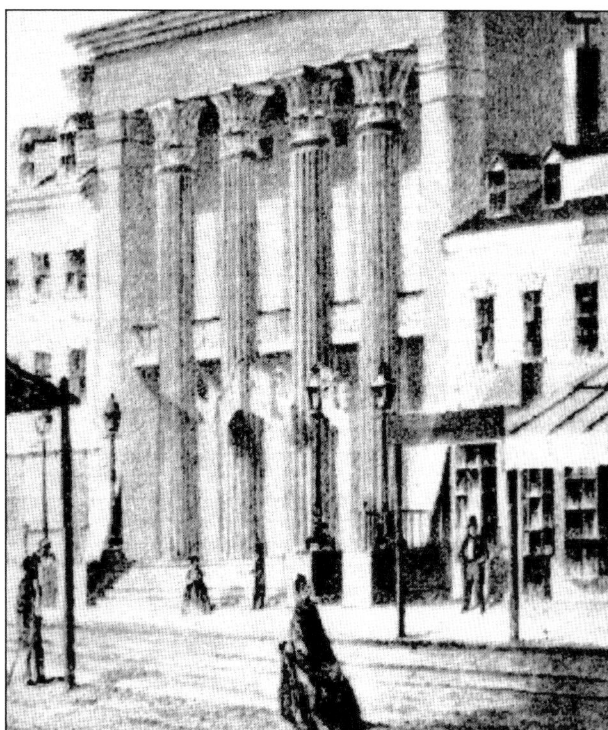

BOWERY THEATRE. It was at this fire that the idea of using wet mats or blankets to protect property was conceived. During the fire "Old Matt" Carey of the Volunteer Fire Department ordered the men to wet carpets and mats and use them to stop the spread of the flames. In 1845, it was adopted as the fire patrol's primary tool for property protection.

New York Board of Fire Insurance Companies,

Monday, April 17, 1865.

At a meeting held this day, at the Insurance Rooms, No. 156 Broadway, Messrs. Edward A. Stansbury, Robert O. Glover and James M. McLean were appointed a Committee to report Resolutions expressive of the sense of this Board, in view of the recent bereavement of the Nation, who reported the following Preamble and Resolutions, which were unanimously adopted, and ordered to be published:

Whereas, In the hour of the Nation's joy and exultation at the Victories which promised once more to bring Peace and Union to our distracted country, it has pleased Almighty God to permit the hand of an assassin to strike down the Chief Magistrate, whose wisdom, fidelity and fortitude have guided us through the terrible struggle of the past four years; therefore, be it

Resolved, That the Members of this Board join with the Nation and the whole civilized world in execration of the spirit which has prompted this deed, and in profound and personal sorrow for the bereavement which it has inflicted upon us.

Resolved, That in the presence of this terrible crime, which is but a natural expression of that bitter malignity with which the Rebellion has been conducted from its inception, it would be a mockery to expect the Nation, standing over the fresh grave of its noble, faithful and forgiving Chief, to consent to strike hands with the bloody traitors whose instrument the assassin was, and permit them again to walk unscathed in the land which they have thus smitten anew.

Resolved, That in the eyes of men, as they move through our streets, Slavery and Treason can read the doom that awaits them, and that the time has come when every Loyal man must draw clearly the line between those who stand by the country at all hazards, and those who palter with Treason or sympathise with our enemies.

Resolved, That as an expression of the feeling of this Board, it be recommended to the Companies composing the same, to close their Offices at Twelve o'clock to-morrow, and that they be closed entirely on Wednesday, the day appointed for the funeral of President Lincoln.

GEO. T. HOPE,

President.

Henry A. Oakley,

Secretary, pro tem.

A Nation Mourns, 1865. This proclamation was issued to the New York Board of Fire Underwriters and the fire patrol members after President Lincoln was assassinated on Friday, April 14, 1865. All of the patrol houses were draped in black bunting as were most public buildings in the nation. The obvious anger and sadness that prevailed at the time is quite apparent in the words of this proclamation. (TER.)

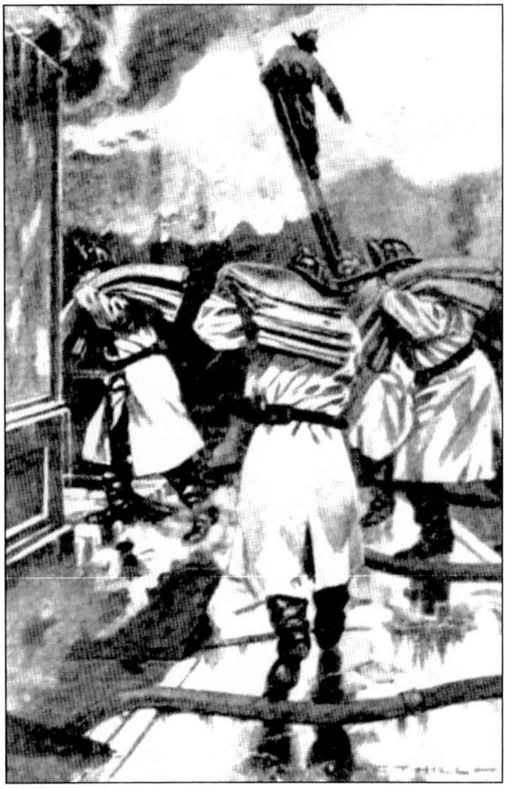

SHOULDERING THE COVERS. This engraving illustrates a scene which has changed little since 1845. The patrolmen in their long white rubber coats have each shouldered two covers and are heading into the blaze. Quickly throwing the canvas covers over stock and goods, the men will endeavor to minimize loss from smoke and water damage.

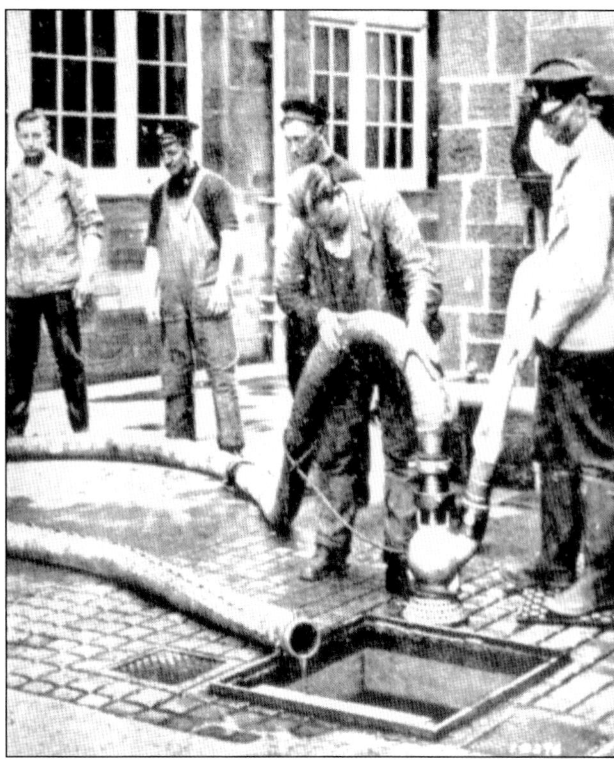

PUMPING OUT. As early as 1850, the patrol realized the need for appliances to pump water out of cellars and basements. Water that cannot drain from buildings must be pumped out as it can cause additional fires and weaken foundations leading to building collapses.

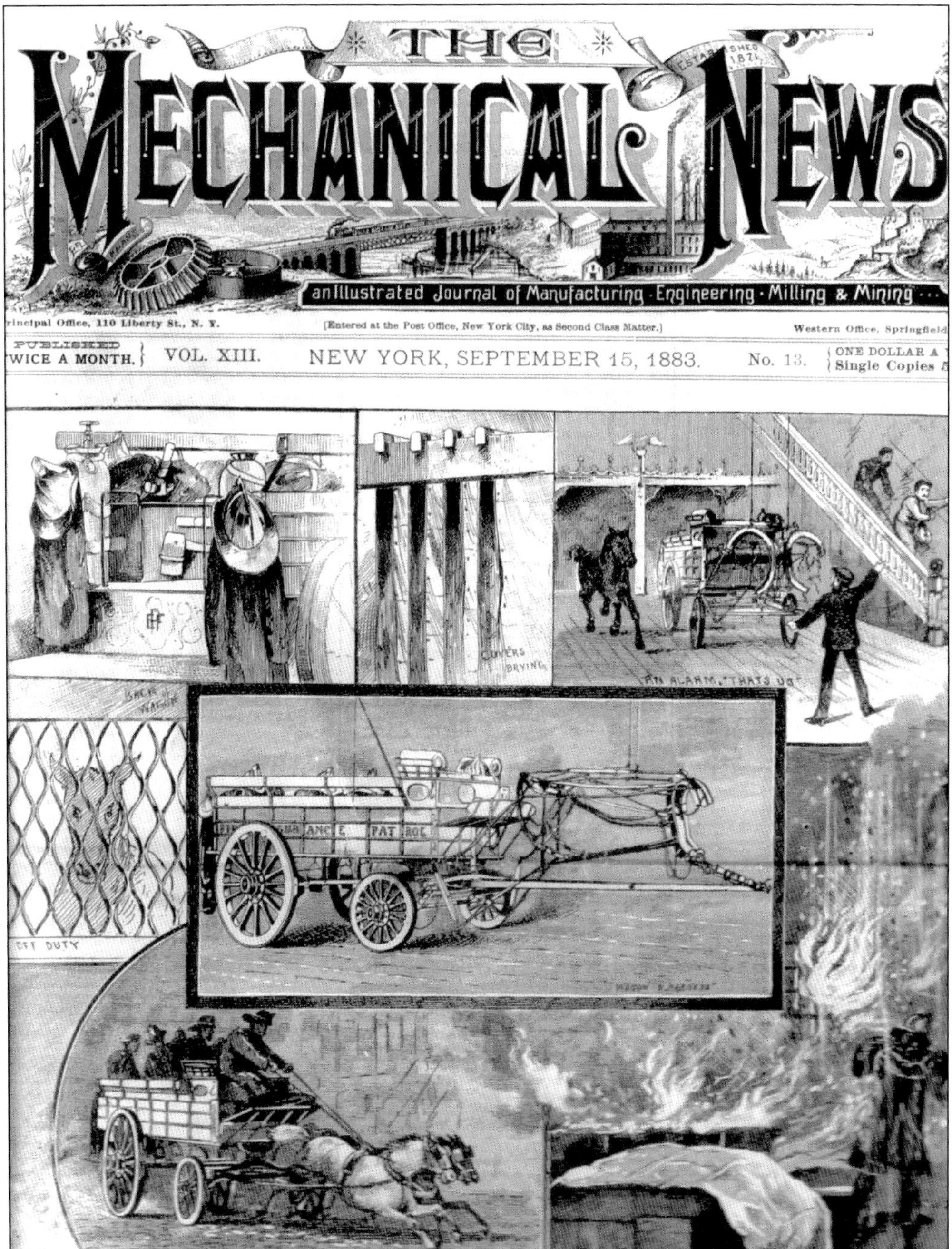

The Life of the Patrol, 1883. In September of 1883, the *Mechanics News* featured this engraving of the Insurance Patrol's daily life. The vignettes are entitled clockwise "Back of Wagon," "Covers Drying," "An Alarm that's Us," "Covering Goods," "At Work," "Off Duty," and "Wagon and Harness" in the center. At this time, the patrol was responding from four locations and had recently changed each company name from their location designation (upper, lower, central, and eastside) to that of a number designation (Patrol 1, 2, 3, 4). (LW.)

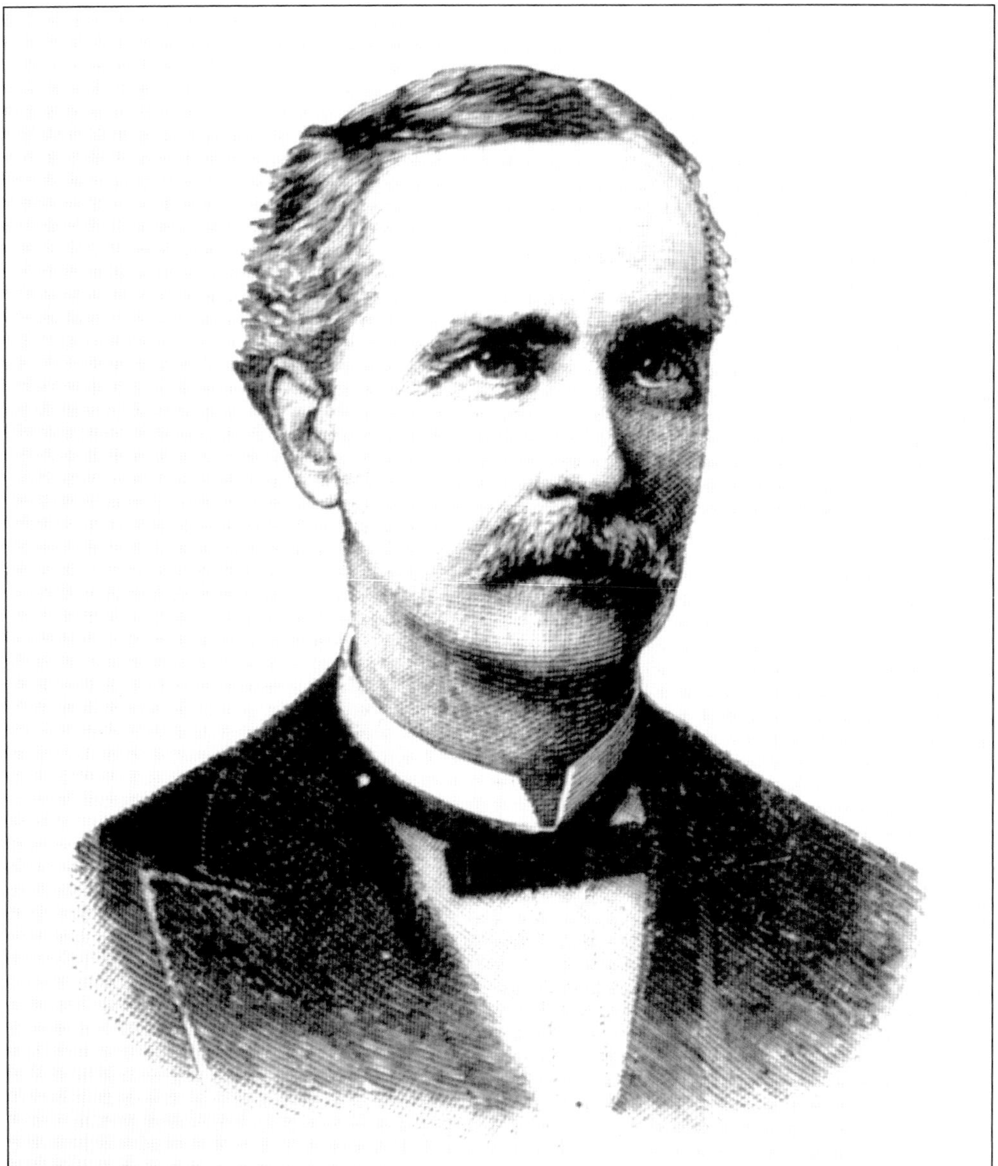

WILLIAM M. RANDELL. William Randell joined the old Volunteer Fire Department in 1847 as a member of Liberty Hose Company 10. He later helped organize Naiad Hose Company 53, which he remained a member of until the company disbanded. He was appointed to the Fire Patrol Committee In 1865 and held the position of secretary for the New York Board of Fire Underwriters (NYBFU) for over 20 years. Randell began compiling information on fire patrol operations and published the first annual report in 1866.

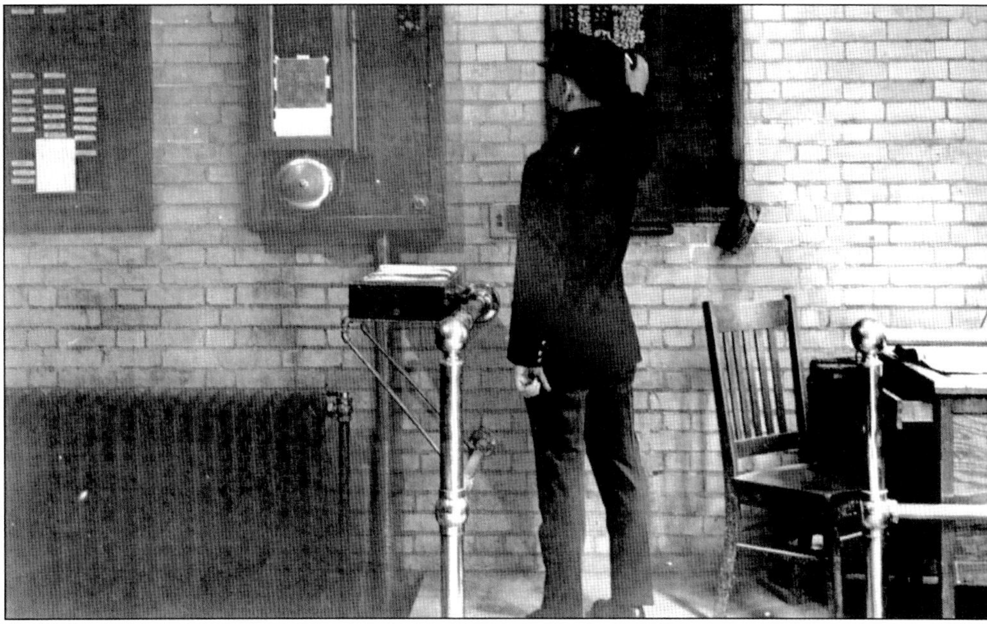

THE HOUSEWATCH. Quickly writing down the box location of the next run is the patrolman at the housewatch of Patrol 1 around 1920. Even as early as the 1850s, a member known as the doorman was assigned to stay behind during alarms. Housewatch duty was and is an important function of the job. Aside from receiving the alarm, counting the bells, opening the doors, and operating the horse harnesses, the patrolman assigned to housewatch duty was charged with logging all persons entering and leaving the building and tending to the coal furnace. (NYFP.)

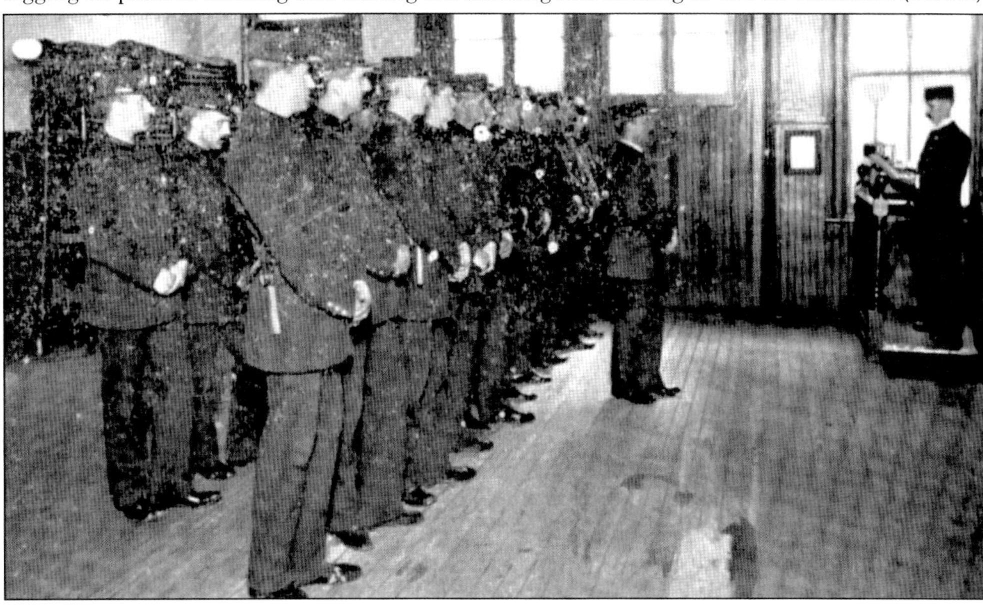

ROLL CALL, 1890. "Patrolman Jones," "Here sir," "Patrolman Tierney," "Here sir . . ." was the familiar sound heard each morning at the 8:00 a.m. roll call. Here members of Patrol 2, at the Old Great Jones Street Station stand at ease, while the captain checks for absent members and reads the day's assignments from the housewatch. Roll call is still held each day at 8:00 a.m., although tool belts and hammers are not required to be worn. (DK.)

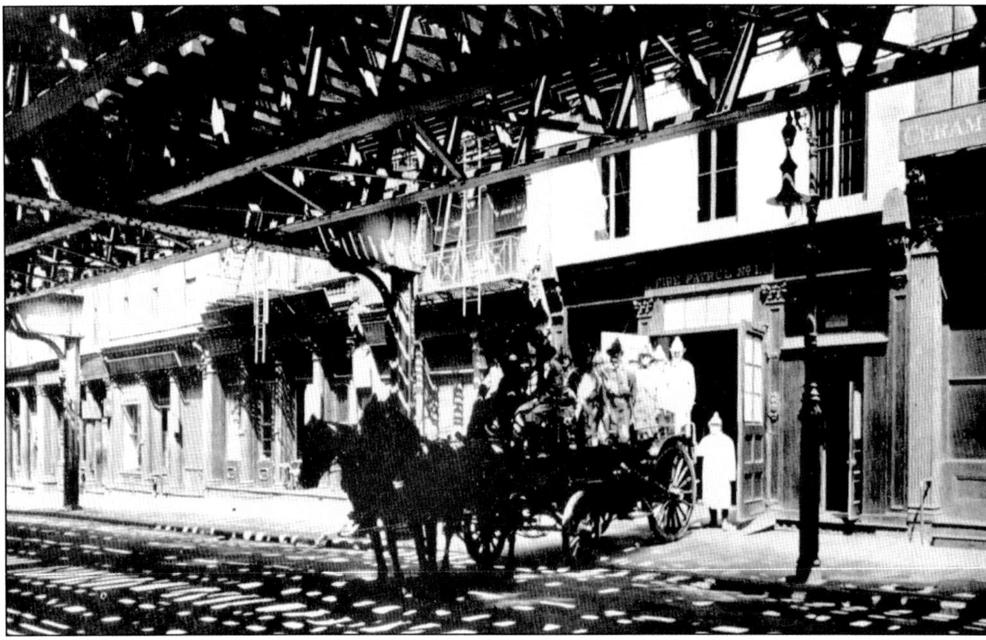

OLD PATROL 1. In 1883, the lower patrol was relocated into a five-story building located at 41 Murray Street in the shadow of the Sixth Avenue El. During this time the rank of sergeant was established, and John Murray was appointed to that position on February 20, 1884. Just four years later, Murray was killed in the line of duty when struck by a falling elevator during a fire at 33 Seventh Avenue. (NYFP.)

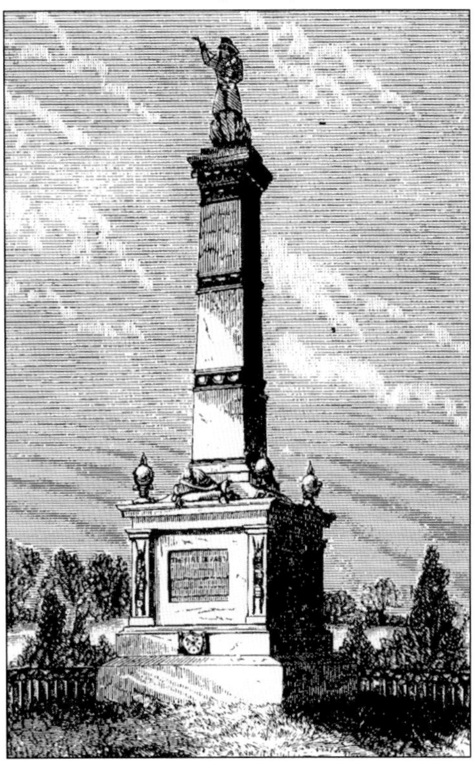

GRAVE OF PATROLMAN CARMAN. Located in Brooklyn's Greenwood Cemetery, this monument was constructed by the Volunteer Fire Department in 1848 as place for all fireman killed in the line of duty to be laid to rest. On October 30, 1853, fire patrolman John S. Carman was killed while battling a large fire on Fulton Street. He was given a full fire department funeral and was buried as a member of Protection (Honey Bee) Engine Company 5 of which he was an active member. He became the first fire patrolman to lose his life in the line of duty. (*Our Fireman.*)

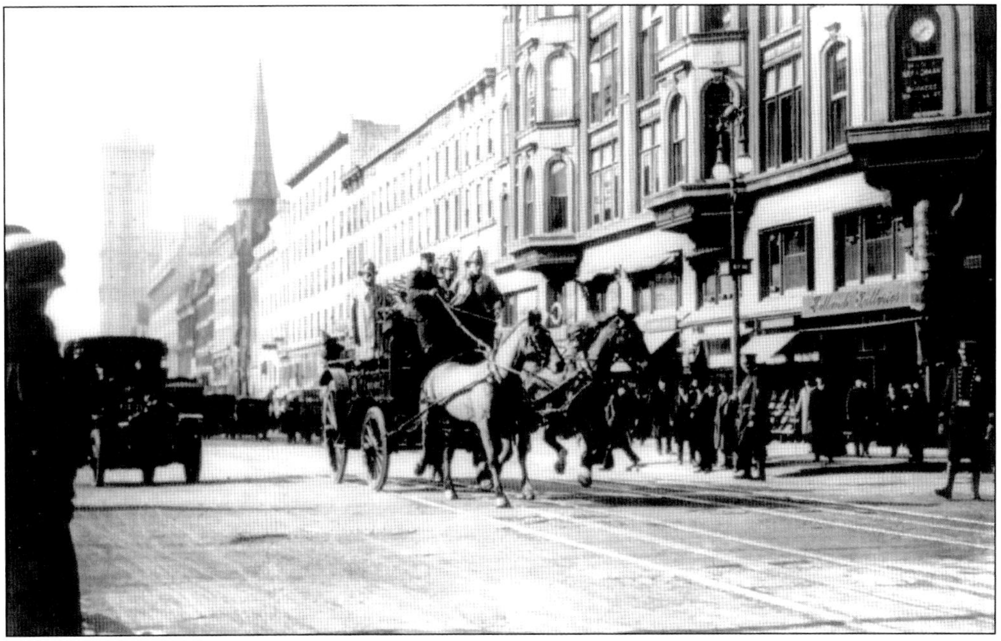

HEADING TO THE TRIANGLE FIRE. One of the worst losses of life to occur in New York City struck on March 25, 1911. Here the second due patrol is rushing to the scene of the Triangle Shirtwaist fire, in which 146 young women perished, almost all garment workers who were trapped on the ninth floor. The patrol went to work that day covering the bodies of the young women, as well as spreading 350 covers in the building. (Library of Congress.)

TRIANGLE FIRE. The Asch Building was completed in 1901 on the corner of Green Street and Washington Place, overlooking Washington Square Park. It was among many of the new "fire proof" buildings constructed during that time in the city. Of course the building was considered to be fire proof due to its expansive use of steel and concrete, however the contents of the building were not "fire proof," evidenced by the fact that 146 employees died from the heavy smoke from burning fabrics. Some of the victims were as young as 15 years old and mostly European immigrants whose tragic deaths paved the way for sweeping federal worker legislation. (NYFP.)

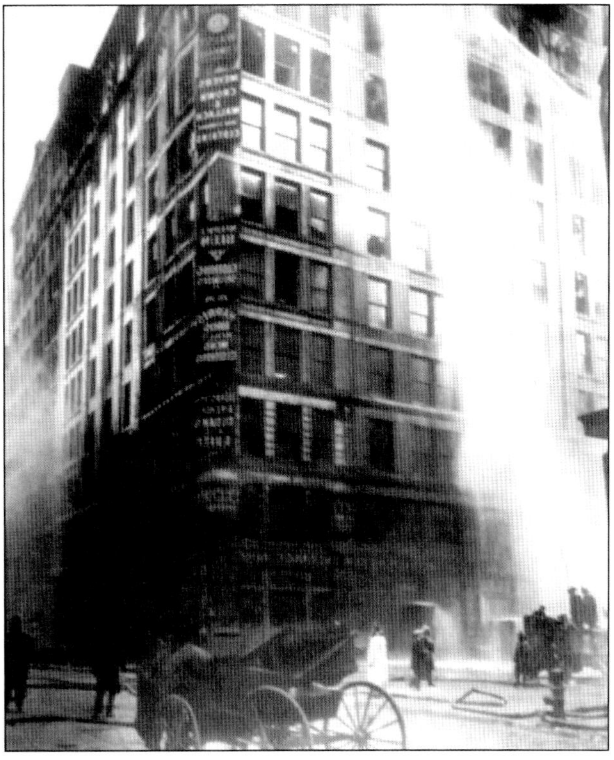

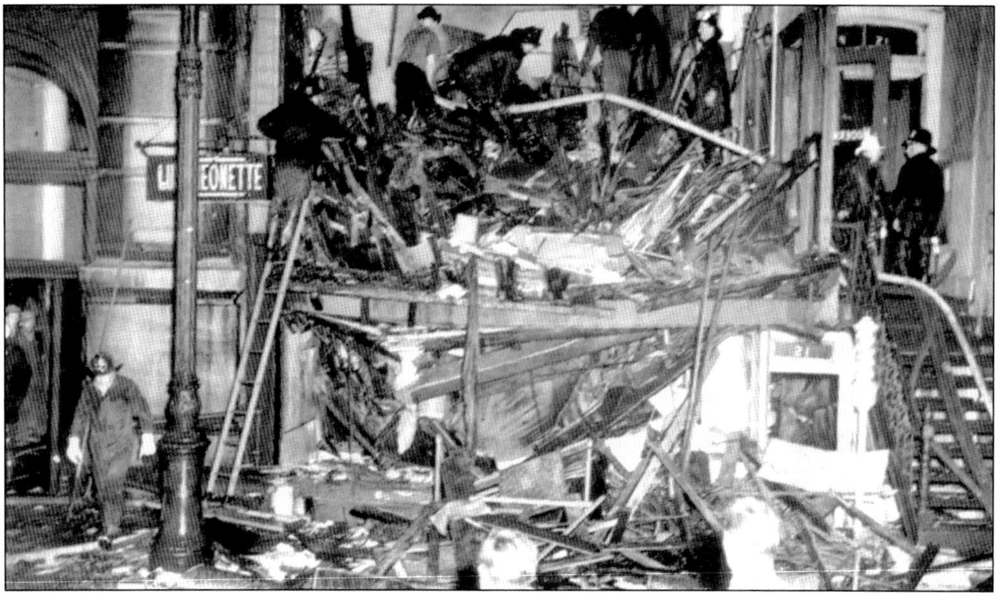

WEST 17TH STREET FIRE. Tragedy struck the fire patrol on October 15, 1949, when this quick moving blaze broke out at 21 West 17th Street. Just after the second alarm had been sounded, the men of Patrol 3 were hard at work in the basement of the building. Soon after arrival, a tremendous collapse occurred of all four floors, burying patrolmen Daniel P. Shea and Frederick Lehman of Patrol 1. These were the NYFP's first line of duty deaths since 1908. (FDNY.)

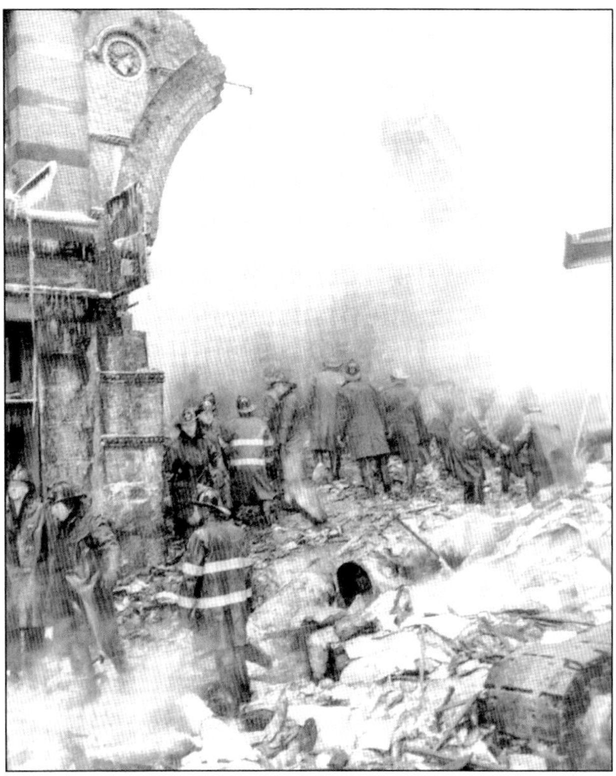

BOX 334, 137 WOOSTER STREET. Disaster struck on Valentines Day in 1958, soon after Patrol 1 responded to a fire at 137–139 Wooster Street. While operating at the third alarm, the top floors of the Elkins Paper Company collapsed without warning, instantly killing the following men; Sgt. Michael McGee, patrolman Louis Brusati, patrolman James Devine, patrolman Michael Tracy, as well as FDNY members Bernard Blumenthal of Ladder 10, and William Schimd of Ladder 1. (FDNY.)

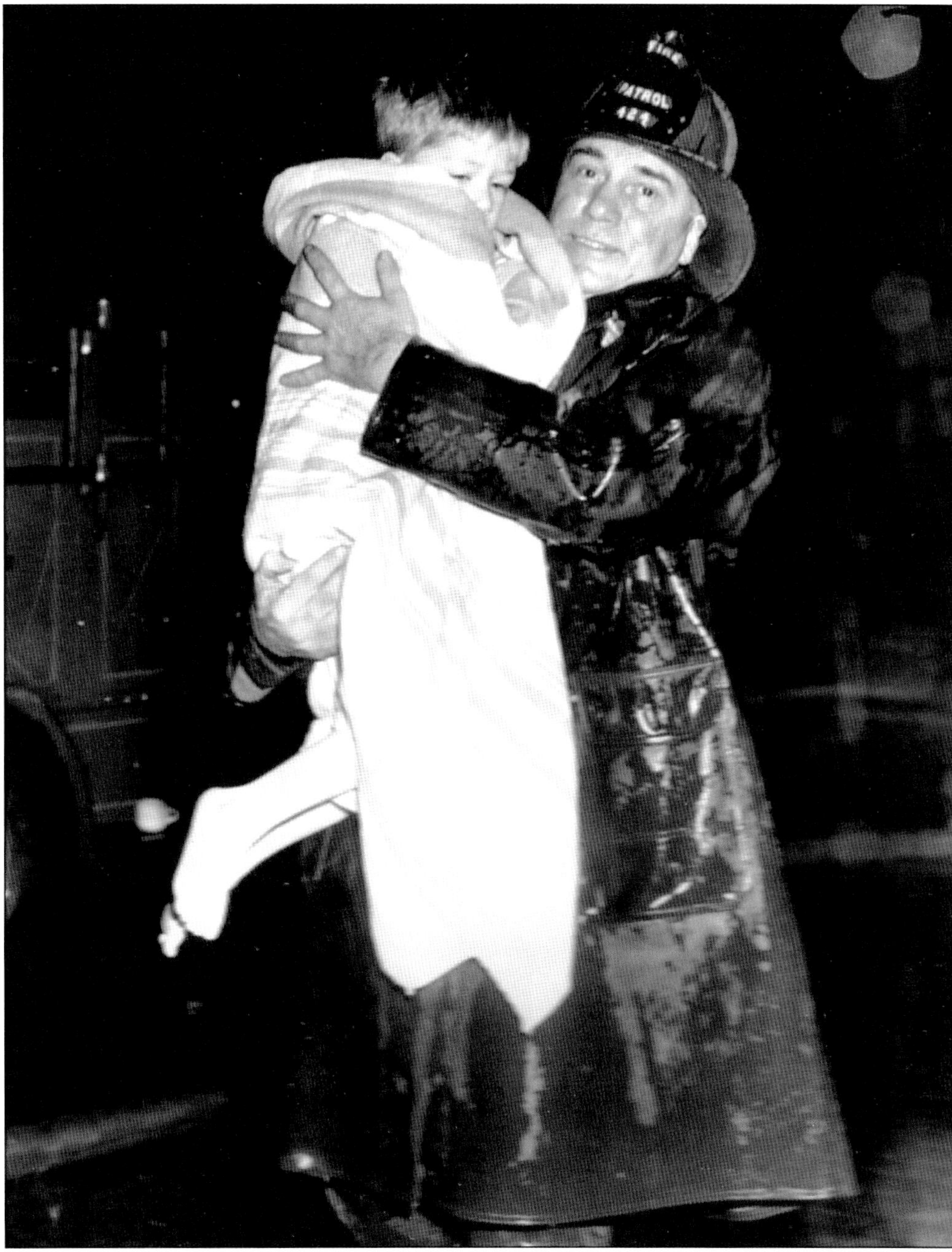

SAVING THE CHILDREN. Throughout the years, the members of the patrol have rescued countless people from fires. In this 1951 Manhattan fifth alarm, the patrol was pressed into service to rescue tenants from the building next door which caught fire just as the last residents had been carried to safety. Here, Raymond W. Cain is seen removing one the last of the children from the building. Patrolman Cain served from December 10, 1930, until he was placed on the inactive list on November 13, 1968. (LM.)

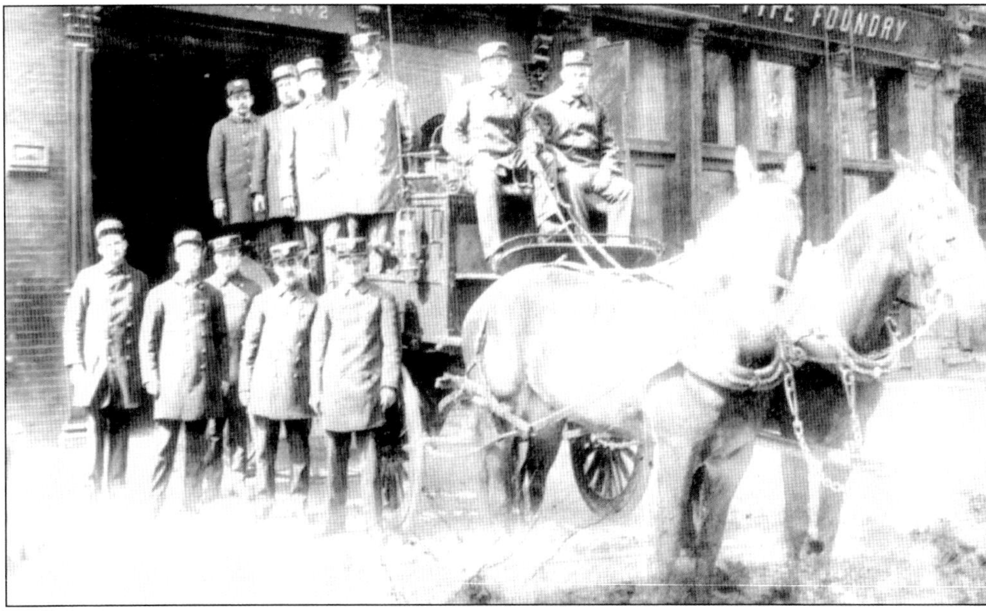

GREAT JONES STREET, 1900. Eleven members of Fire Partol 2 gathered for this photograph, along with two fine fire horses, at their Greenwich Village Headquarters. From 1873 until 1906 Patrol 2 was located here sharing the street with Engine Company 33 who were (and are) located almost directly across the street.

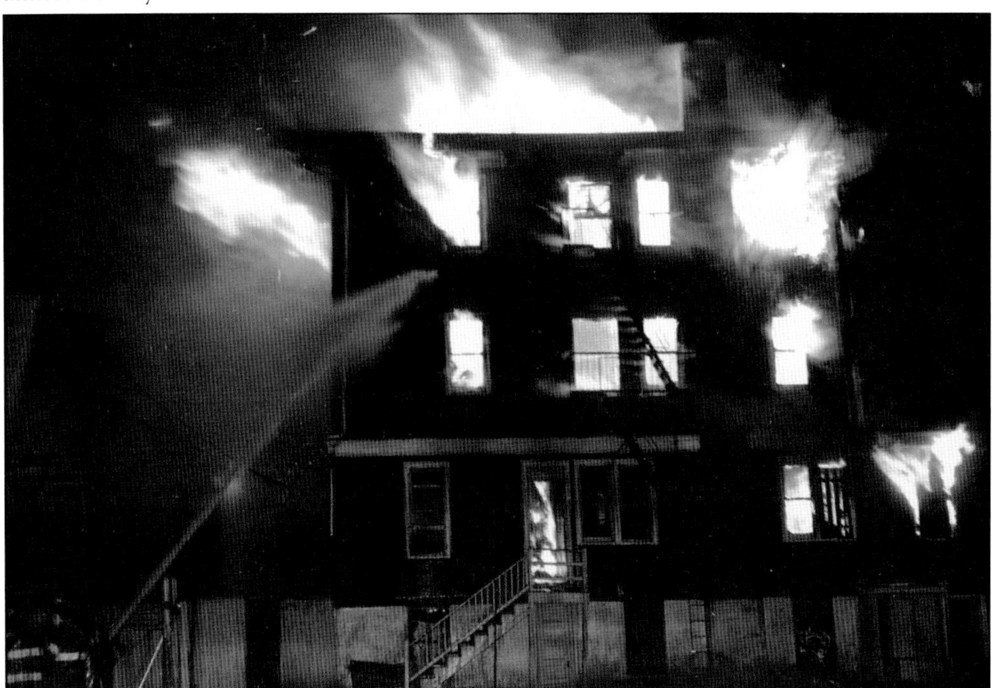

SUMMER HOTEL FIRE. This 1963 five alarm fire destroyed the old Summer Hotel located at 32nd Street and Lewmay Avenue in Queens. Patrol 3 responded to Box 810 that night, and while the hotel was a total loss, much of the exposured property was protected by the patrol's work. (NYFP.)

LUMBERYARD BLAZE, 1950. The patrol responds to all commercial building fires and will often operate at unusual incidents such as this lumberyard fire. Often a few properly placed salvage covers can significantly reduced loss for the property owner and produce savings for the insurance companies. It can also mean the difference to someone's livelihood being destroyed or the ability to open for the business the next day. (NYFP.)

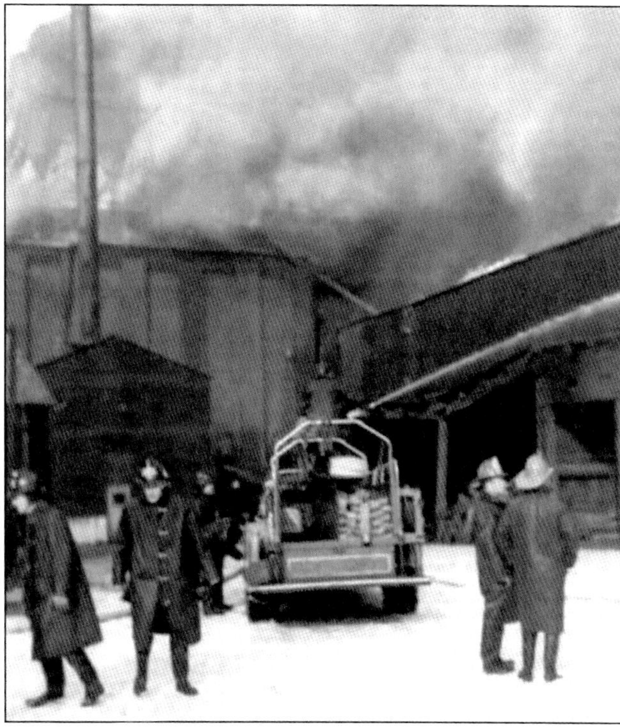

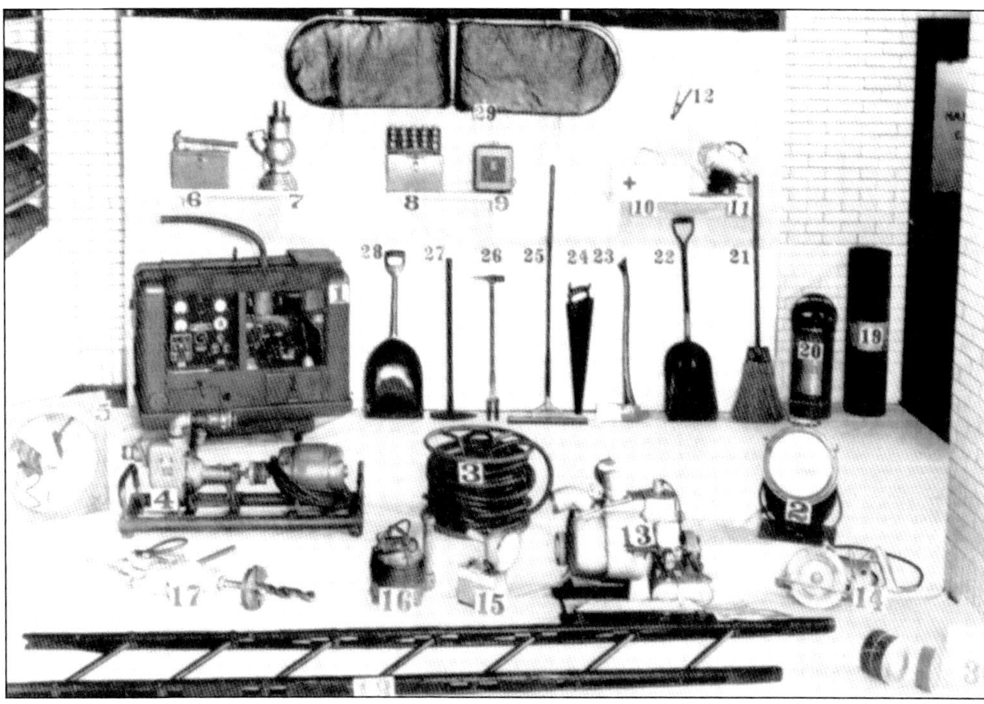

FIRE SALVAGE EQUIPMENT. Fire patrolmen must be jacks-of-all-trades. In the typical day on the job, all of the above tools can be used to save life and property. In any one day, a patrolman might function as a firefighter, sprinkler repairman, elevator operator, plumber, roofer, and many other trades. (NYFP.)

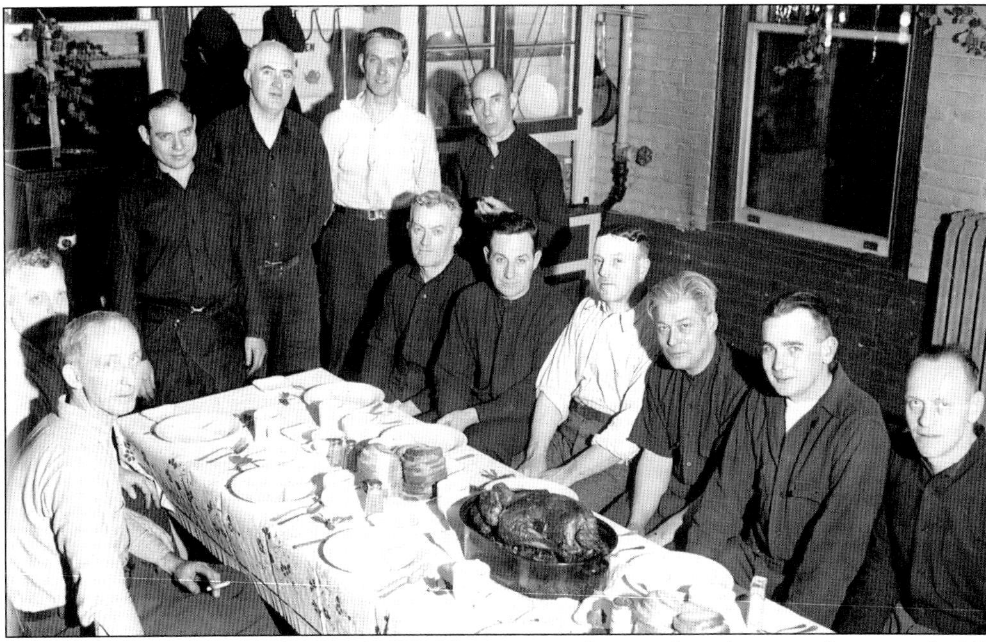

CHRISTMAS, 1940. As with most fire service agencies, the patrol works seven days a week, 365 days a year. These patrolmen made the best of Christmas dinner in 1940 at Patrol 8 (today Patrol 3) at 12 Dean Street. The detail of 12 men were split into two sections, both of which were responsible for answering alarms in the Boroughs of Brooklyn and Queens. Hopefully there were plenty of side dishes, as it looks as though there is only one turkey! (JK.)

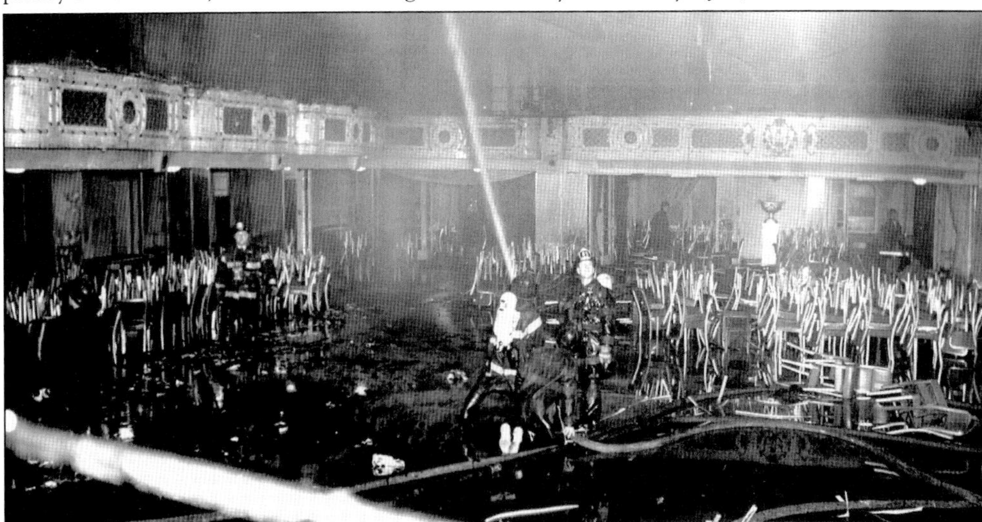

HOTEL ASTOR FIRE. Manhattan Box 811 was transmitted at midnight on January 6, 1958, for fire in the Hotel Astor at 45th Street and Broadway. The detail of Patrol 1 responded and arrived about 70 seconds after the alarm had been transmitted. Under the direction of Chief Henry McKinney, the patrol spread 50 covers and directed water out of the kitchen and protected the expensive appliances with covers. To the hotel management's amazement, the kitchen was able to open for business at 6:00 a.m. that same morning. This is now the site of the MTV Studios. (NYFP.)

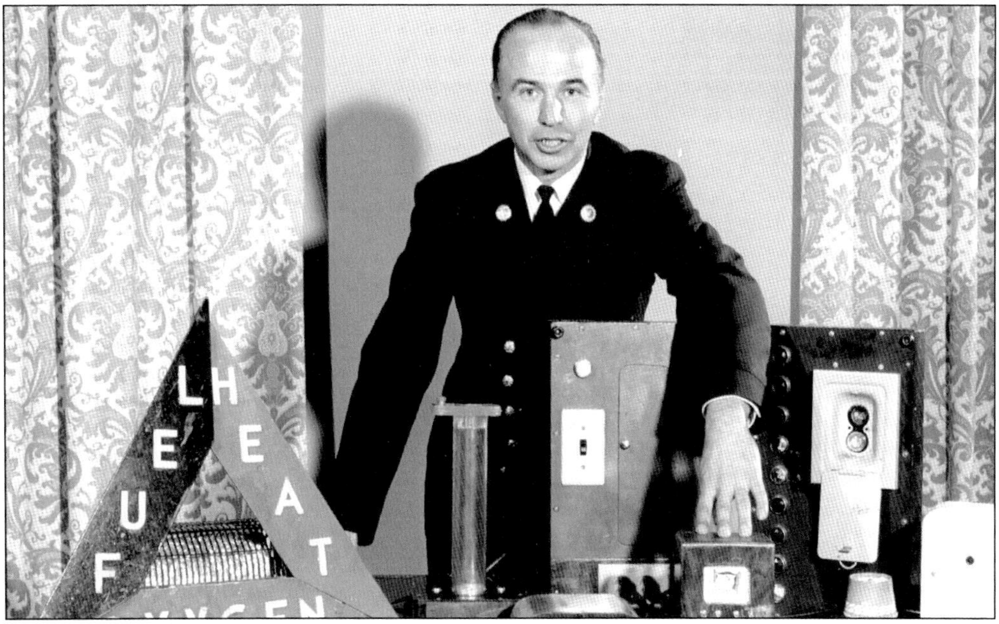

FIRE PREVENTION PROGRAM. The first organized fire prevention program began in New York City in 1920 by members of the patrol. In the early 1930s, fire patrol members constructed this portable fire safety demonstration for city schools. Demonstrating the fire prevention display is Ted Forsberg, a second generation patrolman who was assigned to fire prevention duties on December 27, 1963. (NYFP.)

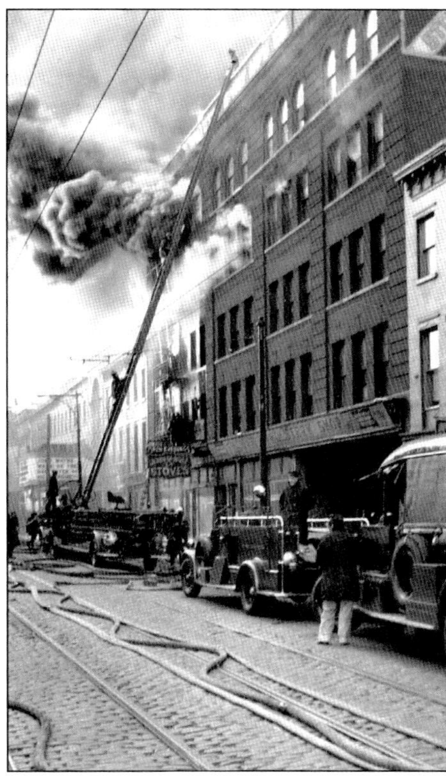

BROOKLYN WAREHOUSE FIRE. This fast moving fifth floor blaze broke out between 1939 and 1954, as evidenced by the Mack rescue truck to the far left. The patrol's rig is near the center of the photograph, with chauffeur apparently awaiting members to return to the truck for more covers. In some instances, the patrol chauffeur will enter the building and assist with throwing covers. This fire is typical of patrol work, although today, most of large multi-story factories have been converted to high end loft residencies. (NYFP.)

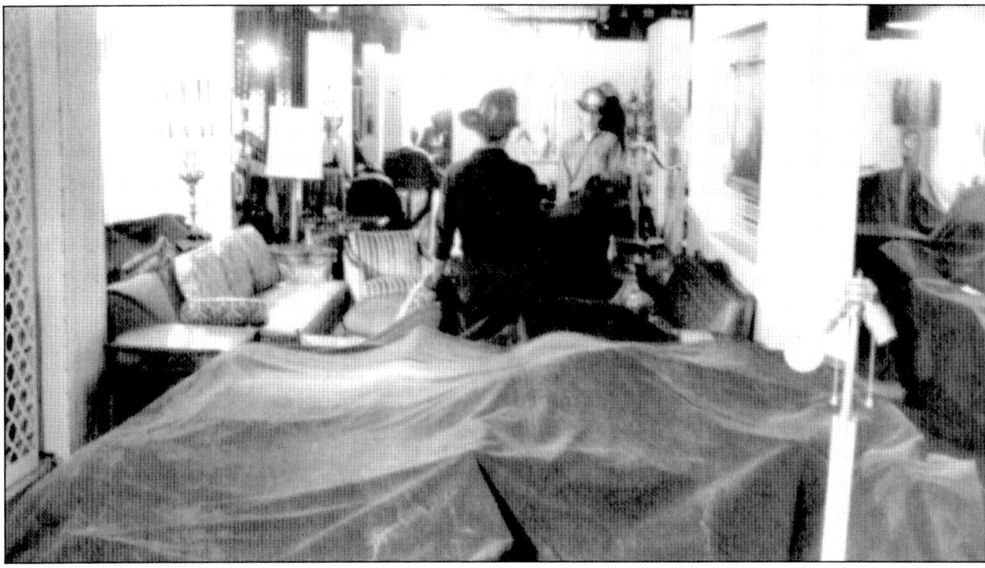

STRIPPING COVERS, 1950. When the fire is out and the water and debris have stopped coming down, the patrol's work is far from done. The next day the detail will return to the scene of the fire and remove or "strip" all covers. They are then returned to the firehouse to be washed, hung, and returned to service. (NYFP.)

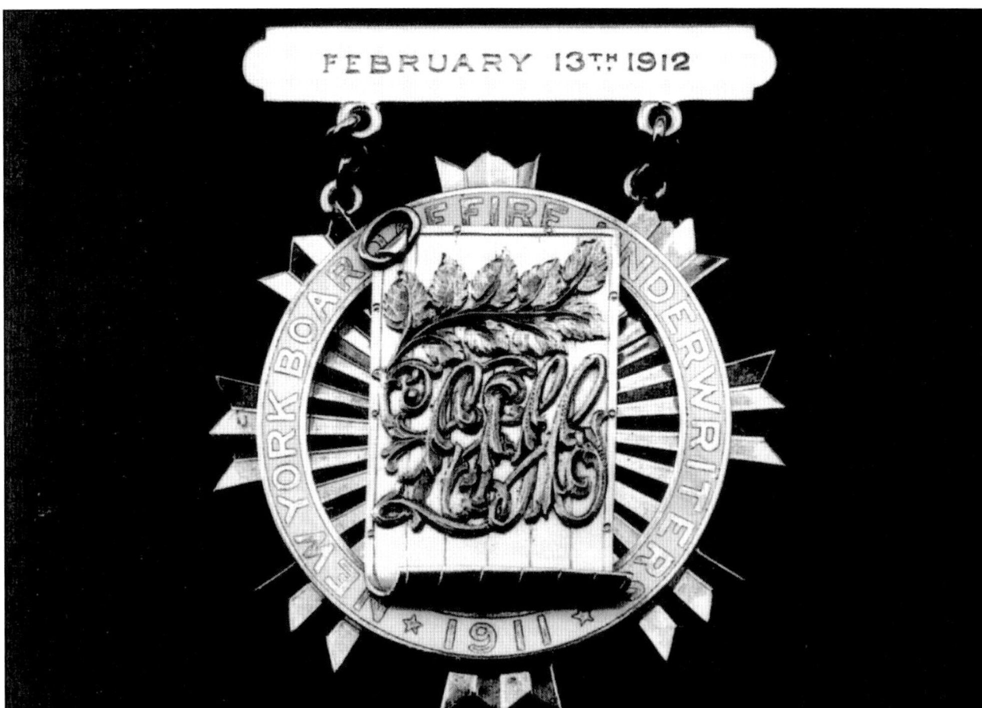

LIFESAVING MEDAL. This medal was awarded to patrolman Louis Haag who was then assigned to Patrol 9, in Brooklyn. On February 7, 1911, patrolman Haag rescued three children and their mother from a third story window during a fire at 26 Powers Street. He later made another rescue in 1912, was promoted to captain in 1932, and died after serving with the patrol for 30 years.

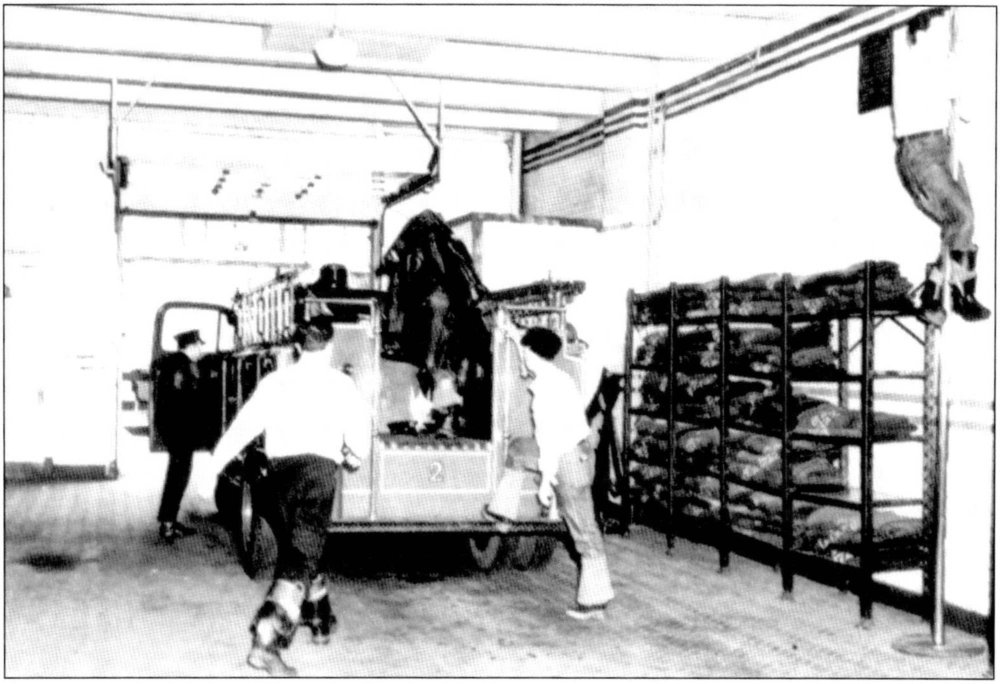

The Patrol on Their Way. Down the pole, into your gear, and out the door is what takes place over 10,000 times a year for the NYFP. Here the detail of Patrol No. 2 is heading out for another fire run in the mid-1950s.

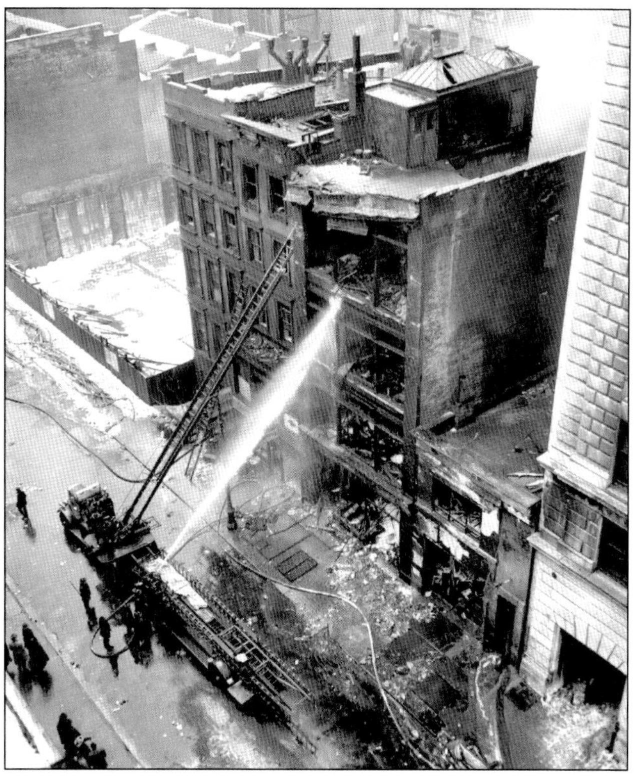

The Aftermath. Many times the patrol will begin salvage operations in a structure, only to realize that the fire is spreading and that all operations must cease. Many times covers and equipment have been destroyed after buildings collapse or become a fully involved fire. Salvage covers are replaced when needed, and about every 10 years all covers are replaced with new ones.

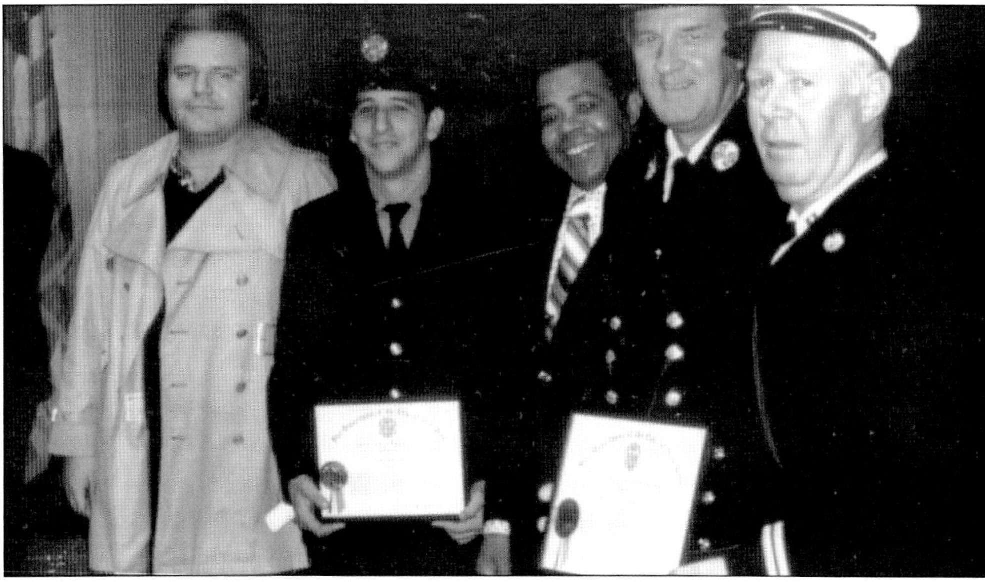

MERITORIOUS SERVICE. On April 12, 1978, Patrol 3 turned out of their Dean Street firehouse to answer a special call for an unusual rescue operation. A worker operating in a street excavation had become trapped in a cave-in; to make matters worse the trench was filling with water, threatening to drown the worker. Patrol 3 arrived on scene, setting up dewatering pumps and succeeded in keeping the rising waters at bay while the FDNY extricated the worker. For their outstanding work the members of the detail were presented with certificates by then FDNY Chief Frank Cruthers. From left to right are patrolman George Munch, patrolman Craig Buccieri, Fire Commissioner Abe Beekman, Chief Cruthers, and Captain Carlson. (JS.)

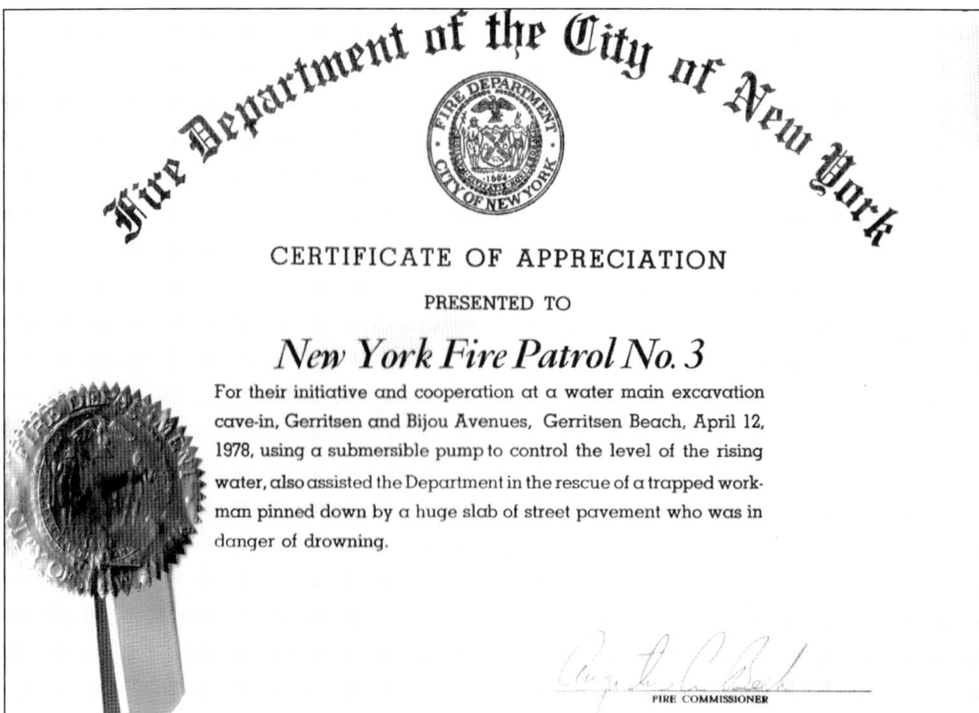

Fire Department of the City of New York

CERTIFICATE OF APPRECIATION

PRESENTED TO

New York Fire Patrol No. 3

For their initiative and cooperation at a water main excavation cave-in, Gerritsen and Bijou Avenues, Gerritsen Beach, April 12, 1978, using a submersible pump to control the level of the rising water, also assisted the Department in the rescue of a trapped workman pinned down by a huge slab of street pavement who was in danger of drowning.

FIRE COMMISSIONER

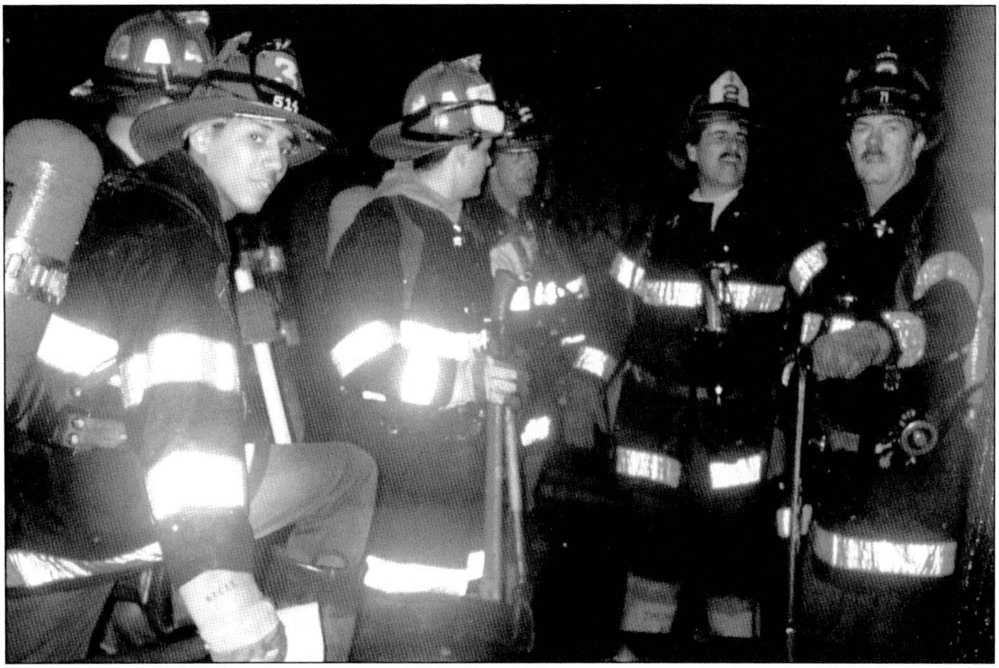

PATROL 2 READY FOR WORK. The members stand at the ready as Sgt. John Sheehan gets ready to enter a Brooklyn fire on Fulton Street. From left to right are patrolmen Gerry Uscatugi, John Ilepi, Bill Tobin, John Sheehan, and Frank Lennon. (FL.)

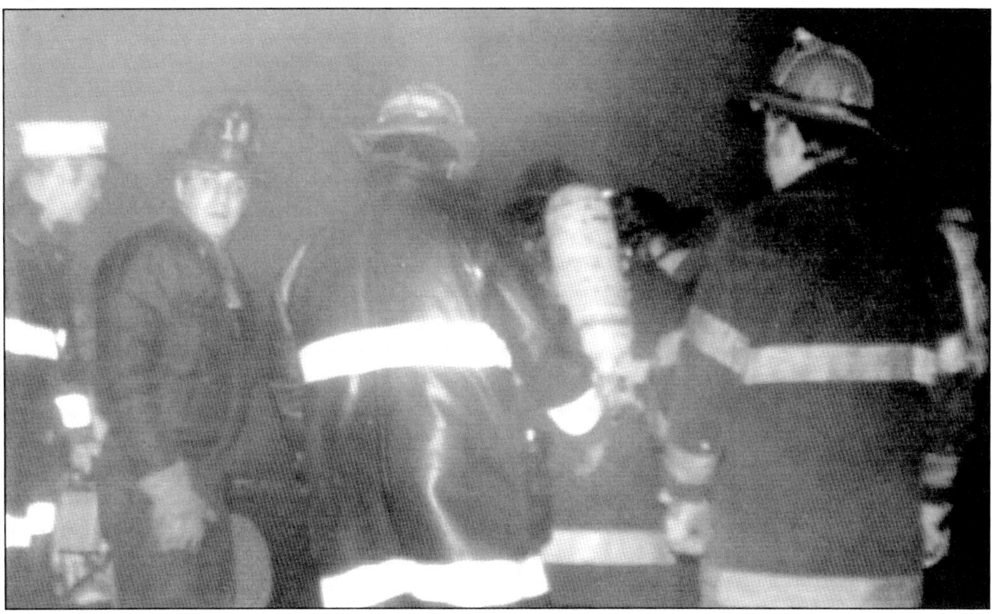

CHANGING BOTTLES AT A FIRE. The smoke is heavy at this job, and the patrol has been pressed into service to change out Scott Air Pak bottles so that the fight against the red devil can continue. The patrol was denied the use of self contained breathing apparatus (SCBA) for many years. However, in 1979, injuries from smoke inhalation forced the adoption of "masks," and in 1980, each company received five MSA SCBAs (mine safety appliance self contained breathing apparatus). (FL.)

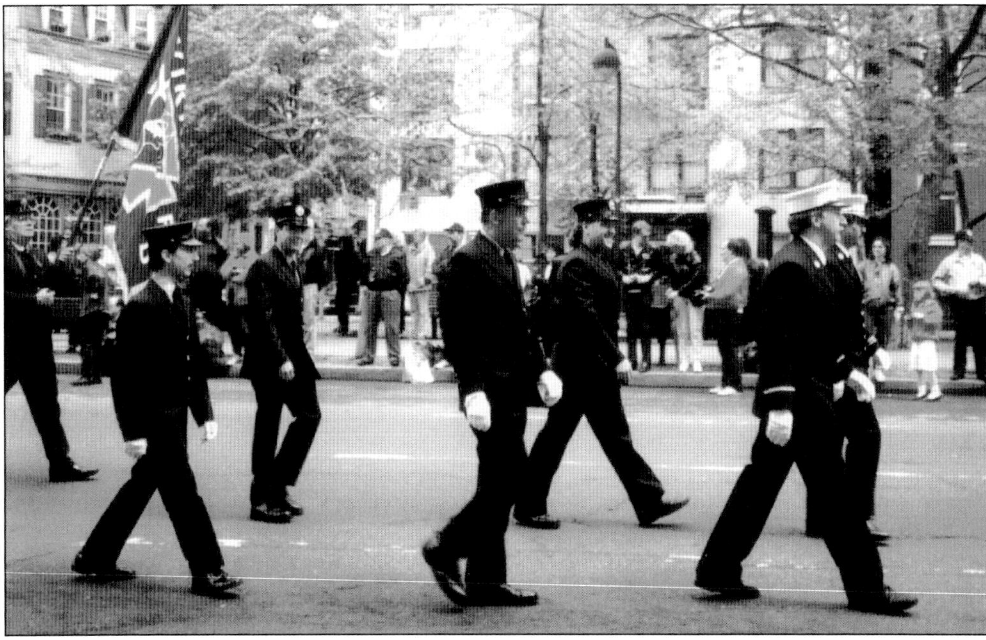

FDNY 125TH ANNIVERSARY. In 1990, the FDNY celebrated their 125th anniversary as a paid fire department. During the parade, members of the patrol proudly marched along the route. Pictured here are some members of the patrol being led by Lt. Anthony Simeoli of No. 2. During the FDNY 125th anniversary, the patrol was celebrating its 151st year of service as a paid unit. (FL.)

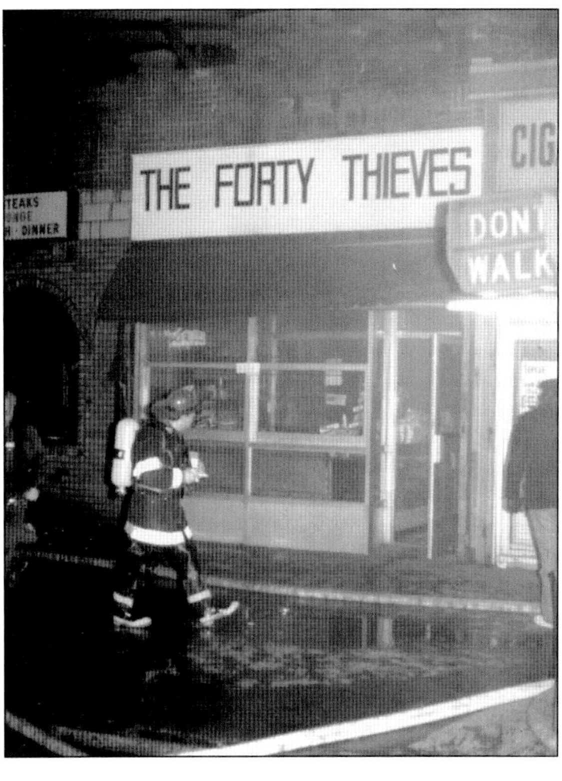

GATHERING INFORMATION. Here Sgt. Charlie "Chuck" Kalfa of the patrol gathers information on the exposures of a fire building in 1976. Information obtained during fires is then entered into what is known as the "bulletin of fires and alarms" which is then distributed to all fire department offices and all insurance companies. (AT.)

ROOF OPERATIONS. Patrolmen must often utilize FDNY aerial apparatus to gain access to roofs. Here the members of Patrol 2 enter a fire escape in downtown Manhattan to gain access to an apartment building. During fires, ventilation operations often cause skylights, roof scuttles, and holes to be opened in roofs. The patrol is charged with covering these openings to prevent the elements from causing further damage to the fire building. Roof covers were first ordered in the 1850s, and in 1959 they were used 179 times. (FL.)

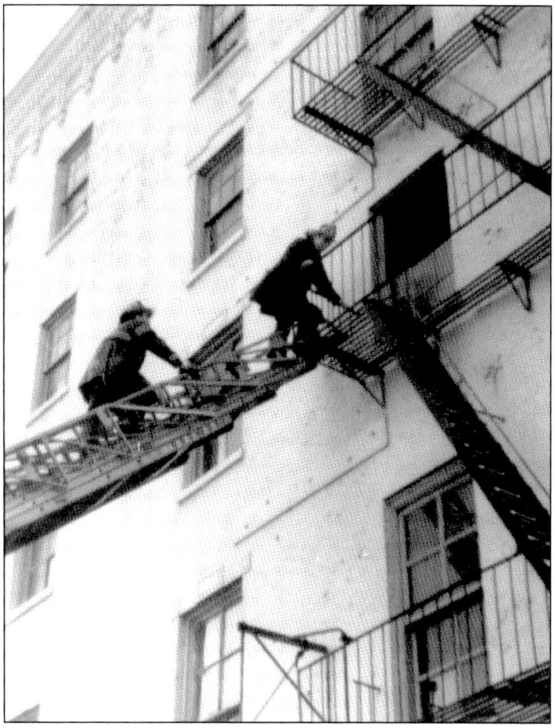

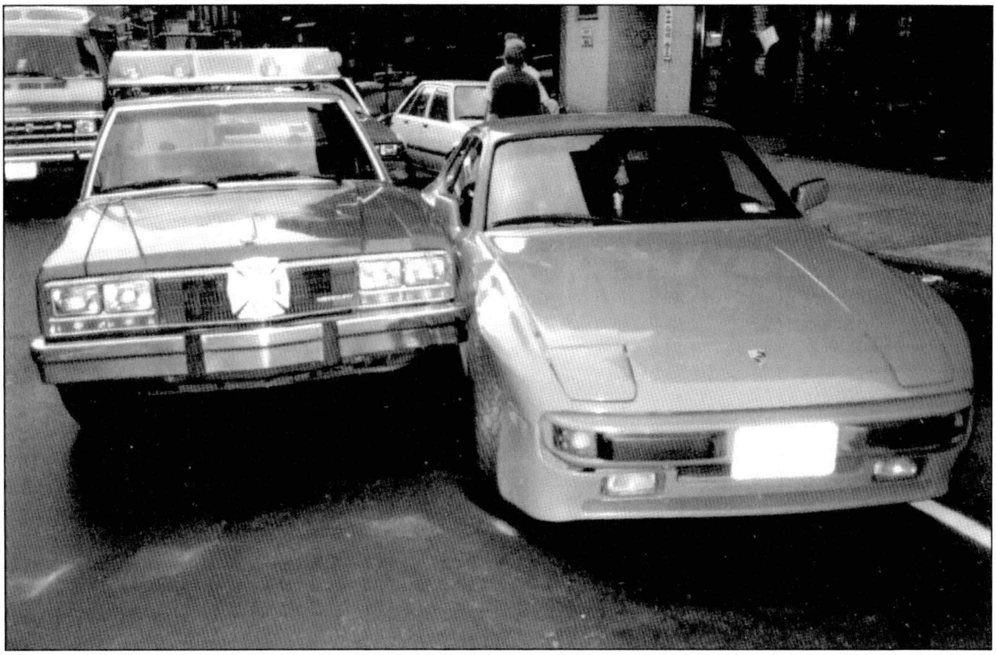

MINOR ACCIDENT FOR CAR 100. It appears that the 1970s Ford Fairmont chief's car will be delayed to the box. This downtown Manhattan crash occurred with a shiny new Porsche. Perhaps since both cars were red, the race car driver thought he had the right-of-way! Emblazoned on the front of the chief's car is the FDNY Maltese cross with the car designation of 100, it should be noted that both the patrol chief and deputy chief vehicles are issued FDNY maltese plates. (NYFP.)

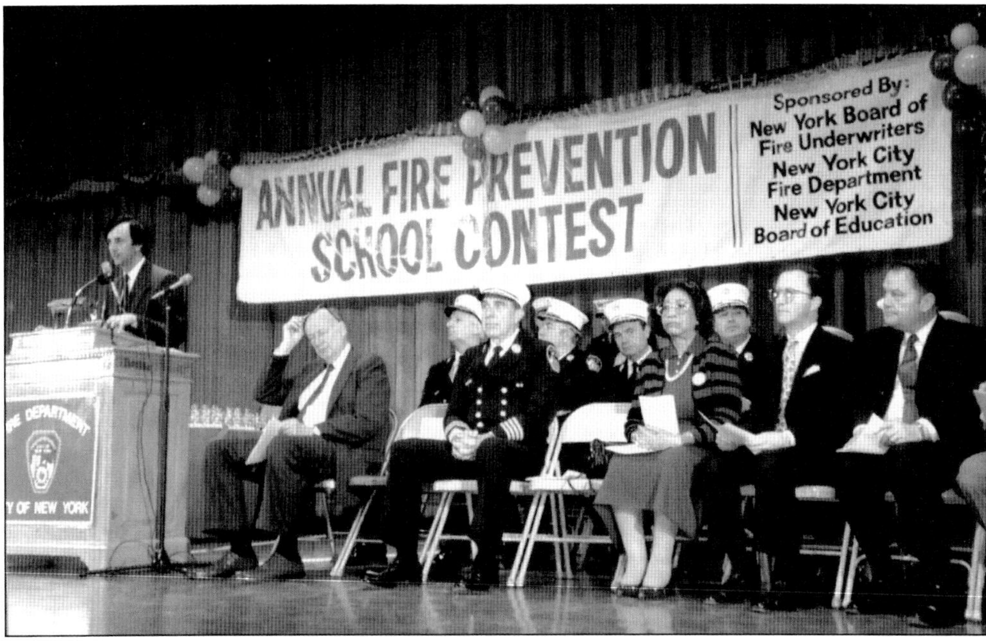

FIRE PREVENTION CONTEST. The fire patrol continues to support the fire prevention campaign in New York City. Here, NYBFU president Alexander Perny, the chief of the fire department, and Mayor Ed Koch attend the ceremony, while fire commissioner Joe Bruno, announces the poster contest winners in 1985. (NYFP.)

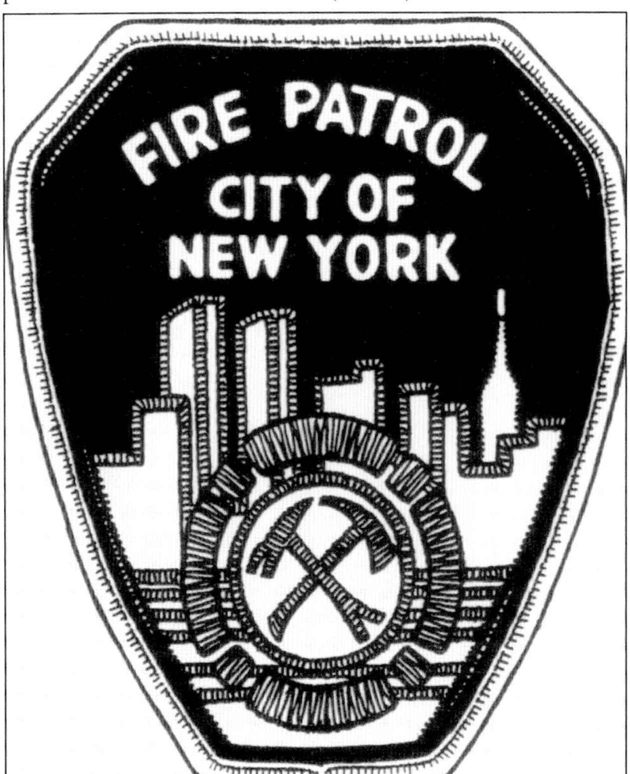

FIRE PATROL EMBLEM. During the late 1970s, the fire patrol adopted their first official shoulder patch. It was patterned off the FDNY patch and contains crossed halligan tool and axe, with the old city skyline in the background. Beginning in 1999, it was first placed on the apparatus and can be found on all equipment as the identifying logo today.

GOING UP. Just like any other city firehouse, local neighbors often ask for help after being locked out of their apartments and homes. The patrol is often on scene with ladders and tools, seen here in Greenwich Village in 1976, climbing ladders and fire escapes and letting people back in, hoping that next time they will leave a key with a trusty neighbor. (FL.)

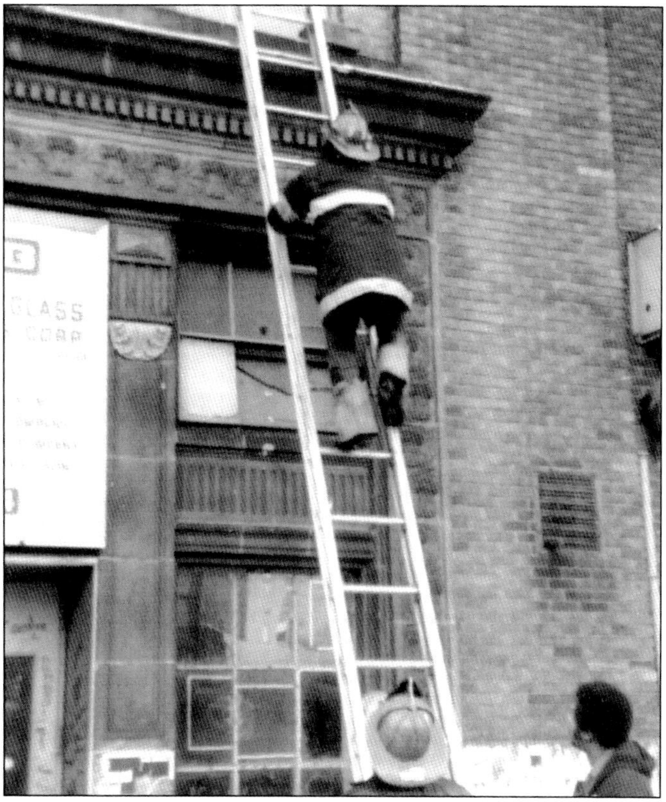

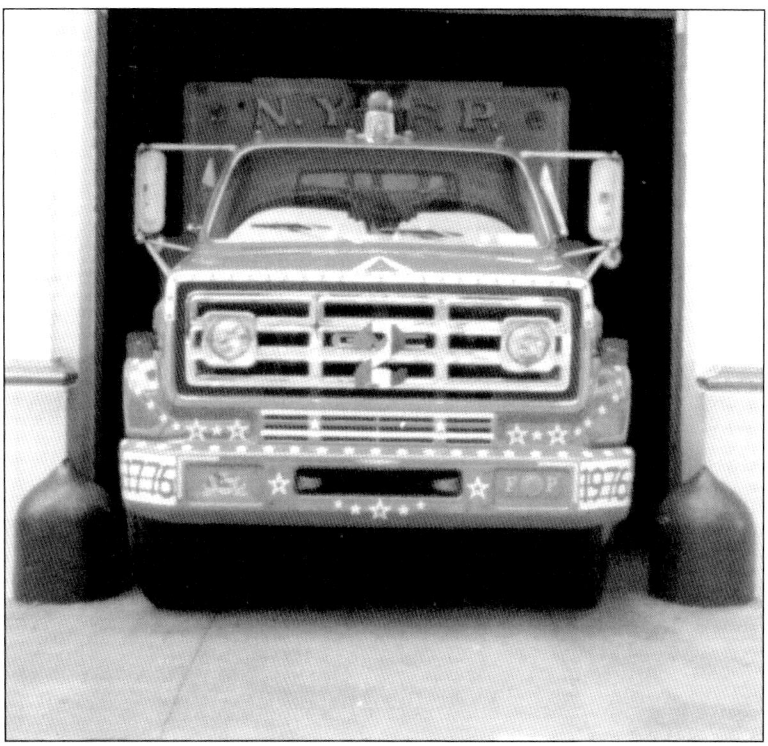

BICENTENNIAL, 1976. There was no stopping the patriotic artwork that appeared on city fire apparatus from 1975 to 1977. Patrol 2 adorned their rigs with stars, stripes and 1776–1976 lettering all across the front and sides of the rig. By 1980, nearly all bicentennial paint schemes had been replaced. (FL.)

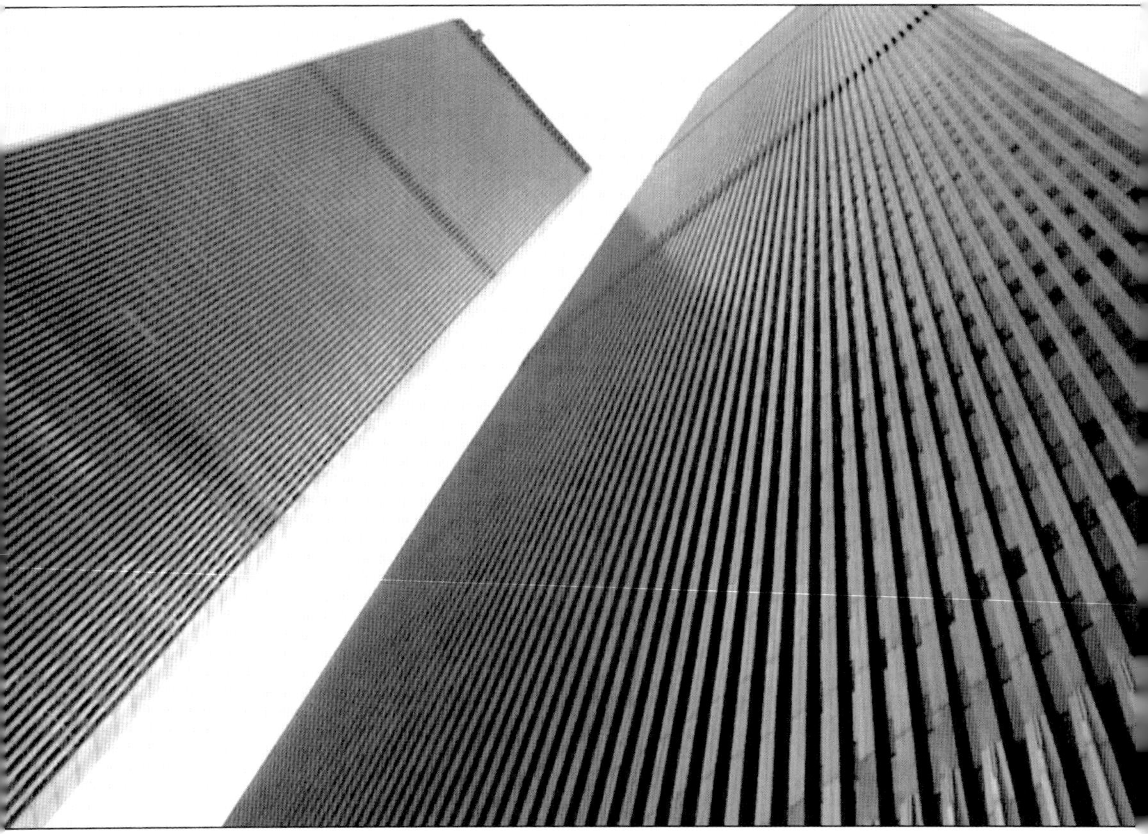

SEPTEMBER 11, 2001. The following members of the patrol reported for duty on that fateful day: Patrol 1, Capt. Tom McGee, patrolmen Robert Renoed, Cal Rolon, Joe Mirro, and Carrel Johnson. Patrol 2 was short staffed with Sgt. Sheehan, patrolmen Joe Krasko, Keith Roma, and Lenny Johnson. Patrol 3 consisted of Sgt. Dennis Kenney, patrolmen Paul Curran, Mike Angelini, John Sottille. Headquarters staff, Chief Ray Carroll, Capt. Peter Greico, and chiefs aide John McGee. Forty-five minutes into the shift, the voice alarm began ringing out as the teleprinter lit up with messages for a plane crash at Manhattan Box 8087. Patrol 2 arrived immediately, followed by the chief and his aide, while Patrol 1 and 3 arrived about six minutes after. Patrol 2's chauffeur, Joe Krasko, was evacuating people fleeing the towers on West Street, while Sgt. Sheehan and Keith Roma were evacuating persons from the South Tower. The first collapse took Roma's life as well as countless others who were killed instantly. The patrol was trying to regroup, trying to find out what had happened, when the second collapse occurred. Only by the grace of God did all escape safely. However, once again, so many others did not.

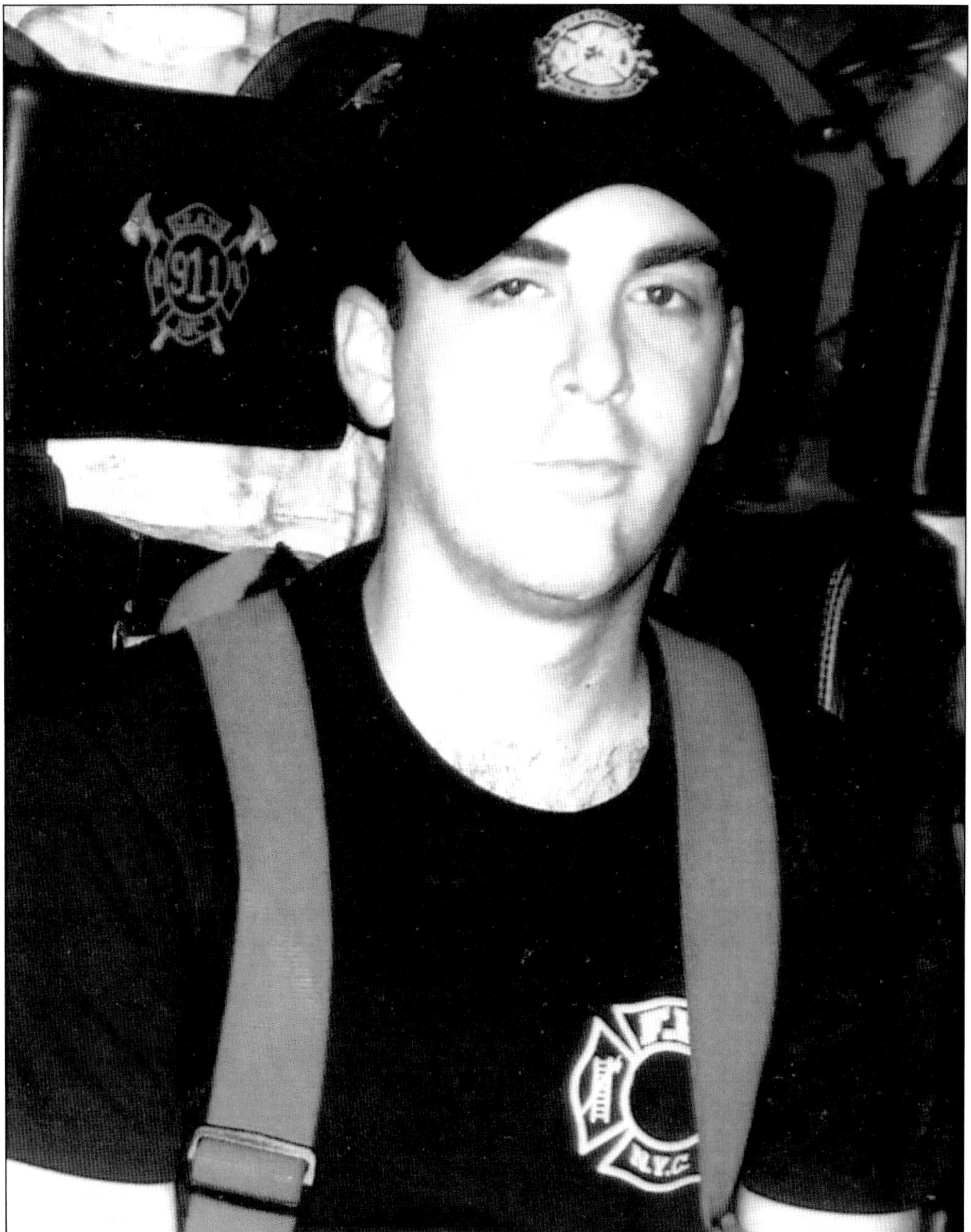

KEITH M. ROMA, 344. Keith Roma was issued badge No. 120 and assigned to Fire Patrol 2 in the heart of Greenwich Village. On the fateful morning of September 11, 2001, when the run came in for Box 8087, Roma was on the rig and ready to go to work as always. A few moments later, the crew of Patrol 2, Sergeant Sheehan and patrolmen Krasko and Roma arrived in front of the World Trade Center. Roma had been evacuating people from the tower for over an hour and, from many eyewitness accounts, he alone removed over 200 people to safety. Patrolmen Keith Roma became the first member of the NYFP to die in the line of duty in over 30 years. He was found on Christmas Eve 2001, near the last group of nine civilians that he was saving; his body lay in front of theirs with his red helmet close to his side. (NYFP.)

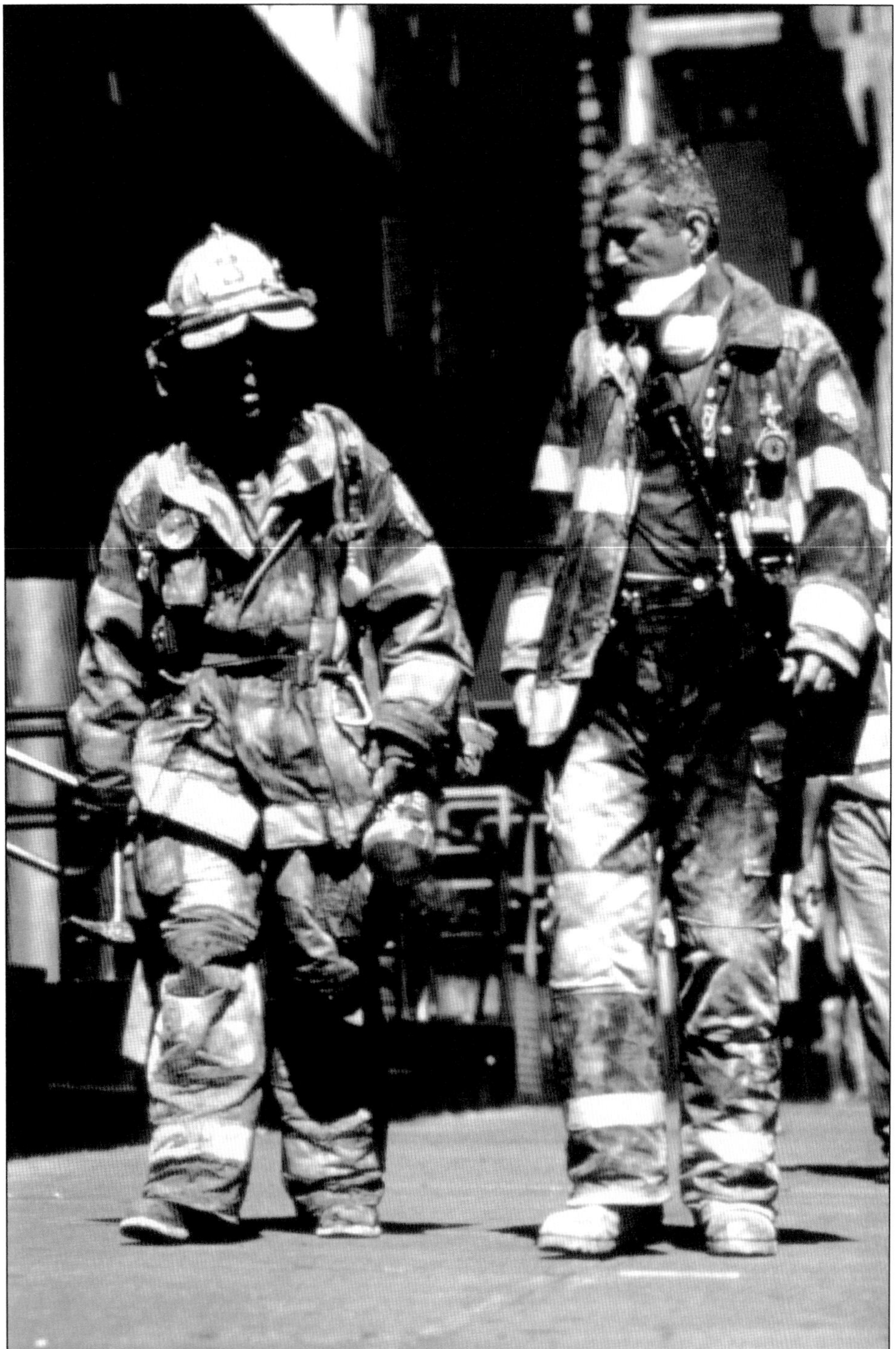

WALKING AWAY. Fire patrol captain Thomas McGee and Sgt. James Canham walk in disbelief after the dust has settled. Both men met up after the collapse and were still searching for other members of the patrol, neither could know what certain death they had escaped. (TM.)

CAPT. PATRICK BROWN. Patrick "Paddy" Brown remains as one of the most well known heroes of the FDNY from over the last three decades. But before his outstanding career began with the FDNY, he was a member of the patrol and spent his days at the busy firehouse of No. 1 on West 13th Street. Captain Brown was killed on September 11, 2001, and was laid to rest by his brothers from Ladder 3. (FL.)

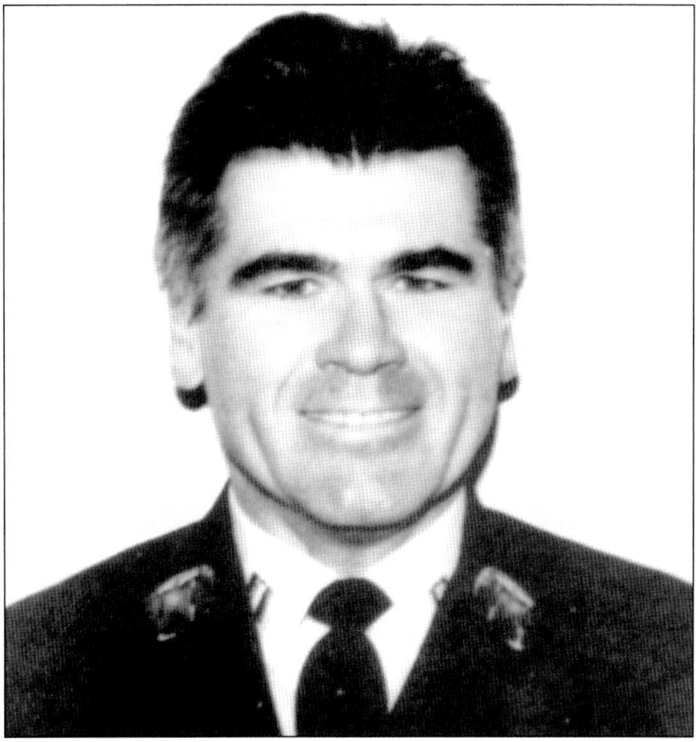

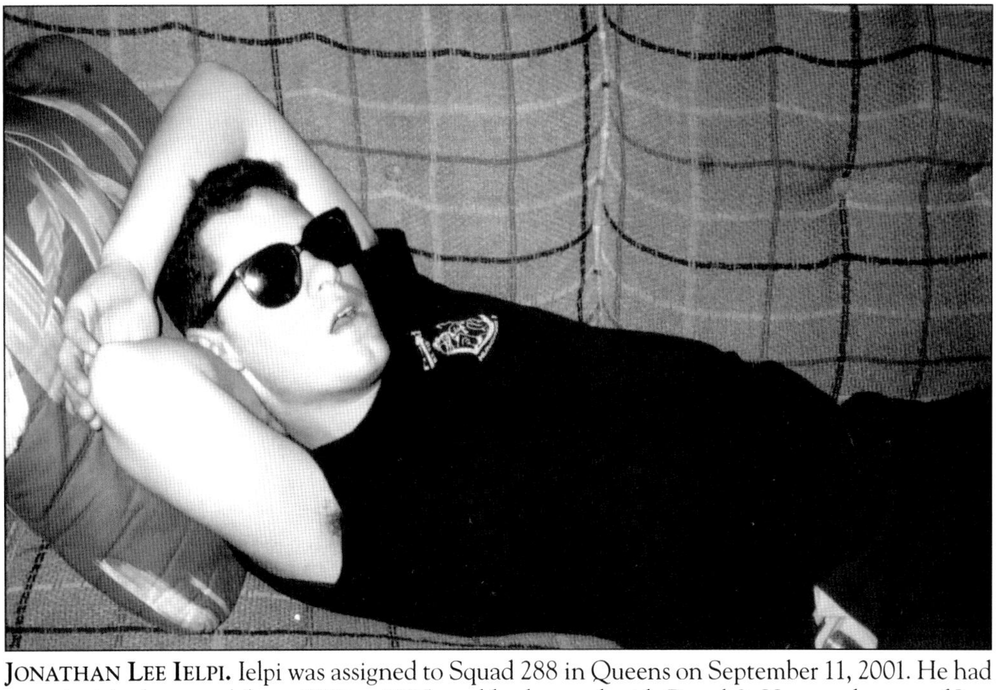

JONATHAN LEE IELPI. Ielpi was assigned to Squad 288 in Queens on September 11, 2001. He had served with the patrol from 1990 to 1995, and had served with Patrol 2. He was the son of Lee Ielpi, a well known and respected member of Rescue 2. Ielpi was admired by all in the NYFP, and will be remember for his great sense of humor and pride. Ielpi was found in 2001 and was buried under the FDNY Honor Guard with his family and comrades nearby. (FL.)

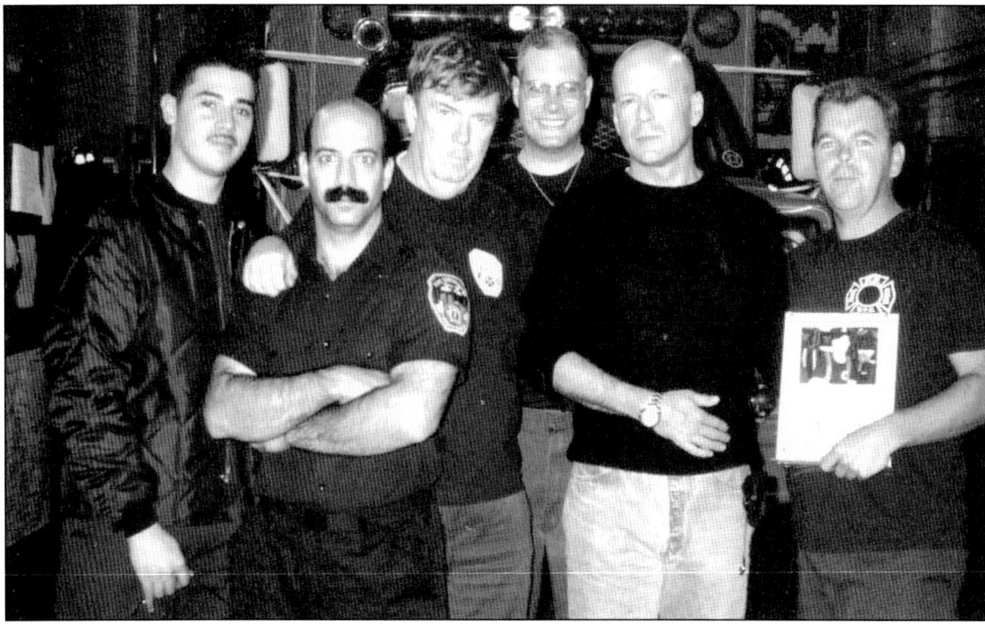

BRUCE WILLIS PAYS HIS RESPECTS. Fire Patrol 2 is in a remarkable location for tourists and celebrities alike. This night, actor Bruce Willis stopped in to pay his respects to Keith Roma's memory shortly after September 11. From left to right are patrolmen Hector Rivera, Anthony "Tony" Emmanuelle, Frank Lennon, Mike Ryan, Bruce Willis, and Mike Farrell who is holding Roma's photograph. (FL.)

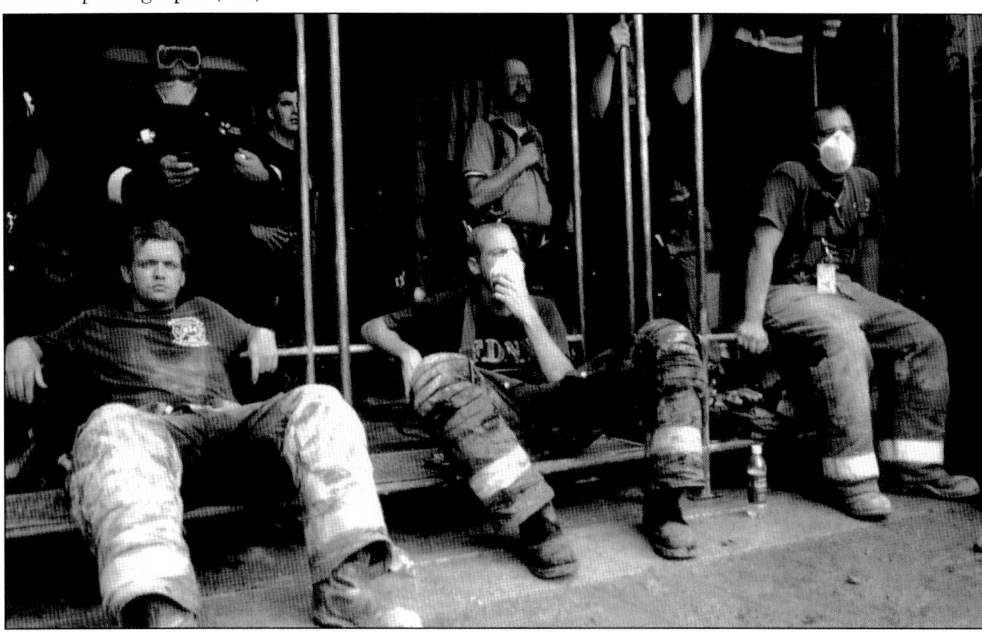

DAILY NEWS PHOTOGRAPH, SEPTEMBER 13, 2001. This photograph shows the destroyed quarters of Ladder 10 and Engine 10 being used as a rest area for firefighters working the pile. Among those in the photograph are fire patrol sergeant Larry Wagonfeld, patrolmen Rich Leonard, Mike Angelini, and Timothy Regan (author). All the members are waiting for another assignment to begin the unthinkable task of finding the survivors that did not exist.

IN MEMORY OF PATROLMAN ROMA. In 2002, this bronze plaque was erected by the officers and members of the fire patrol. Mounted just inside the apparatus door of Patrol 2, it is a fitting tribute to a beloved patrolman, son, and father. Many of thousands of tourists passing through the village stop into the firehouse to read, admire, and often cry at the plaque, which is illuminated 24 hours a day in his memory. (MK.)

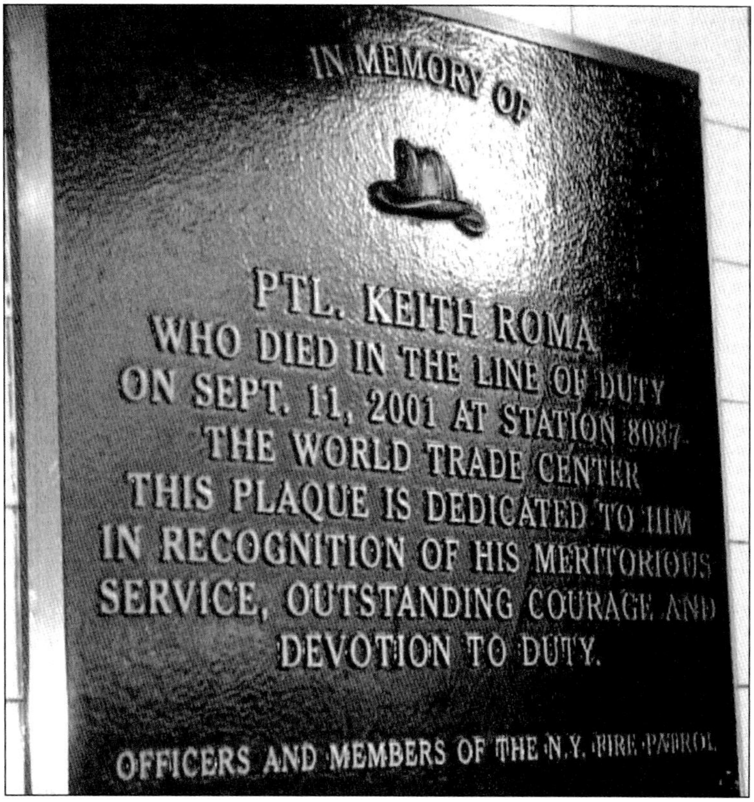

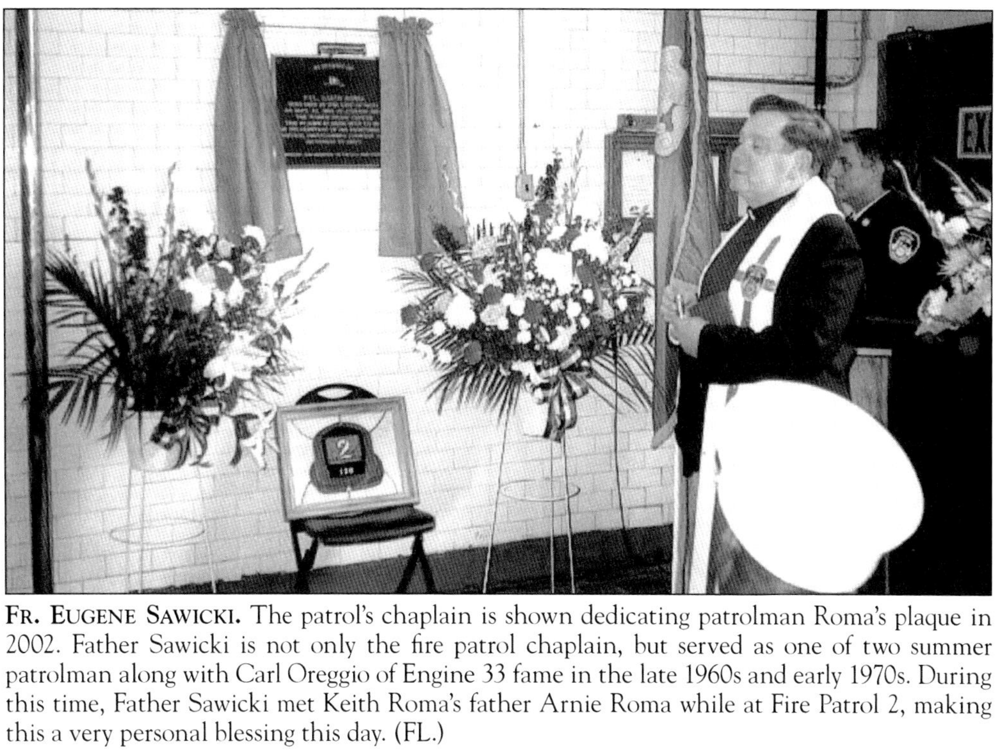

FR. EUGENE SAWICKI. The patrol's chaplain is shown dedicating patrolman Roma's plaque in 2002. Father Sawicki is not only the fire patrol chaplain, but served as one of two summer patrolman along with Carl Oreggio of Engine 33 fame in the late 1960s and early 1970s. During this time, Father Sawicki met Keith Roma's father Arnie Roma while at Fire Patrol 2, making this a very personal blessing this day. (FL.)

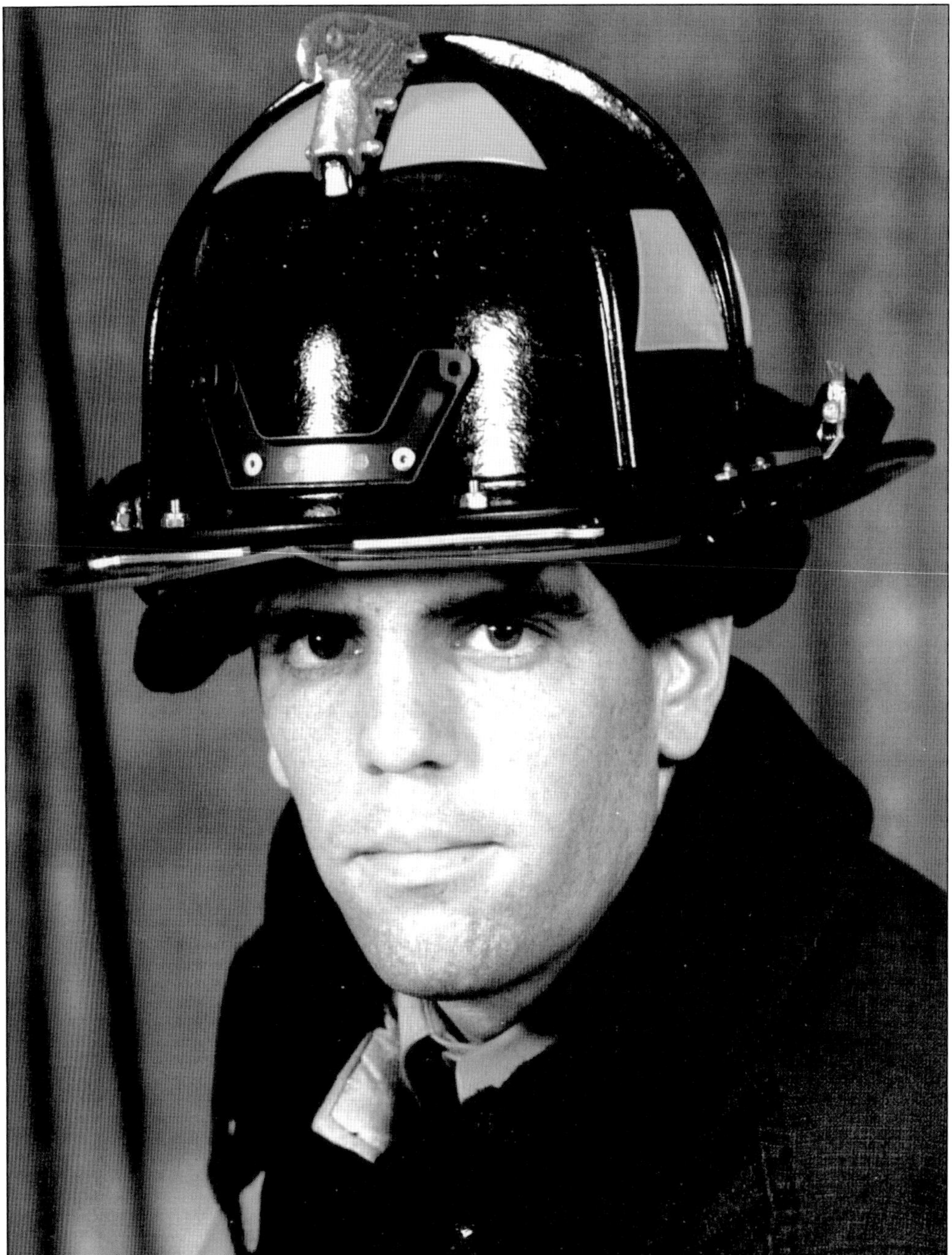

James Raymond Coyle. James "Jimmy" Coyle was assigned to Ladder 3 in Manhattan on September 11, 2001. He had left the fire patrol only a few months earlier. Assigned to Patrol 3, Coyle was the heart of the laughs at the Dean Street firehouse. His dedication carried over to the FDNY where he was buried with honors in October of 2001. In 2003, the City of New York dedicated a section of Avenue R in his Marine Park neighborhood, as James Raymond Coyle Way, in honor of this 26-year-old hero. (NYFP.)

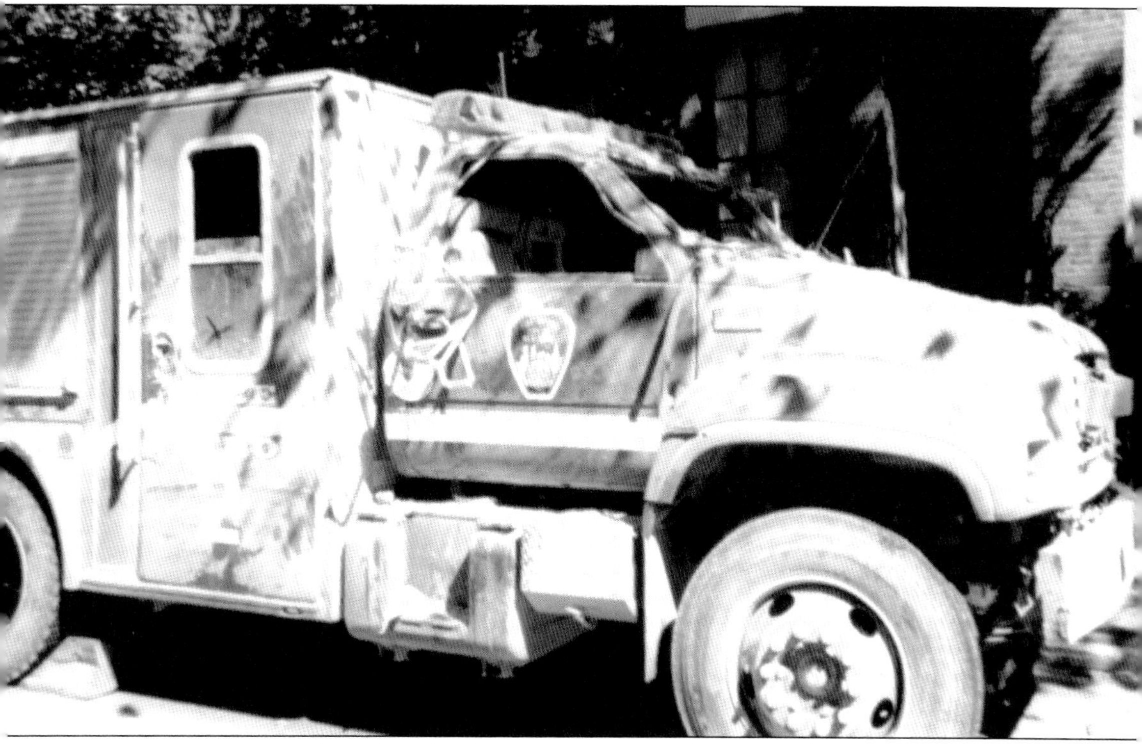

PATROL 2 APPARATUS. The 1999 GMC/Ferrara from Patrol 2 was badly damaged during the collapse. The truck was repaired and is shown here before it was rebuilt in 2002. This apparatus was later dedicated to the memory of Keith Roma. (JK.)

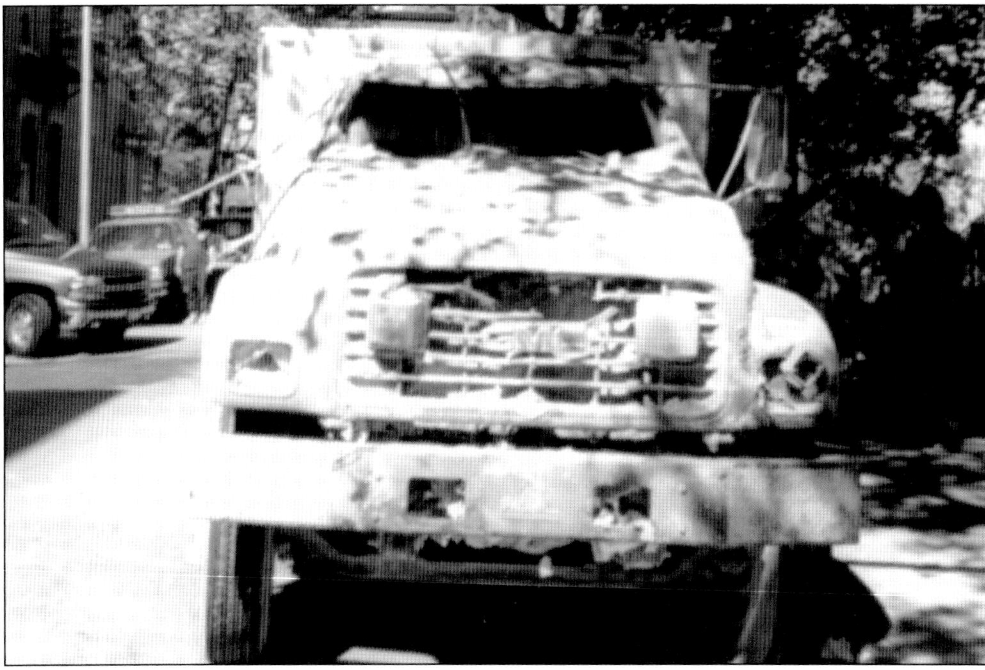

Fire Patrol 2 Apparatus. Here is another view of the destroyed GMC of Patrol 2. The truck was removed from the World Trade Center site with hundreds of pieces of burnt DOS computer manual pages imbedded in the trucks grill and cab. The truck was crushed, burned, and damaged from that fateful day, but was rebuilt soon after by a local truck dealer.

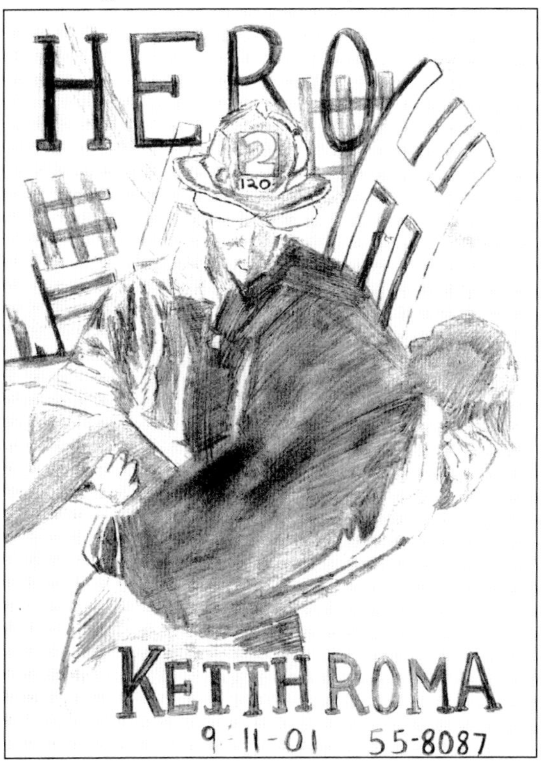

A Tribute to Keith Roma, 2001. The outpouring of public support for the fire service had never been witnessed in New York City before the September 11 attacks. This simple pencil drawing was donated to the members of Patrol 2, in honor of Roma's heroic sacrifice. Badge No. 120 and the name patrolman Keith Roma will never be forgotten in this city.

Two

Patroleo's

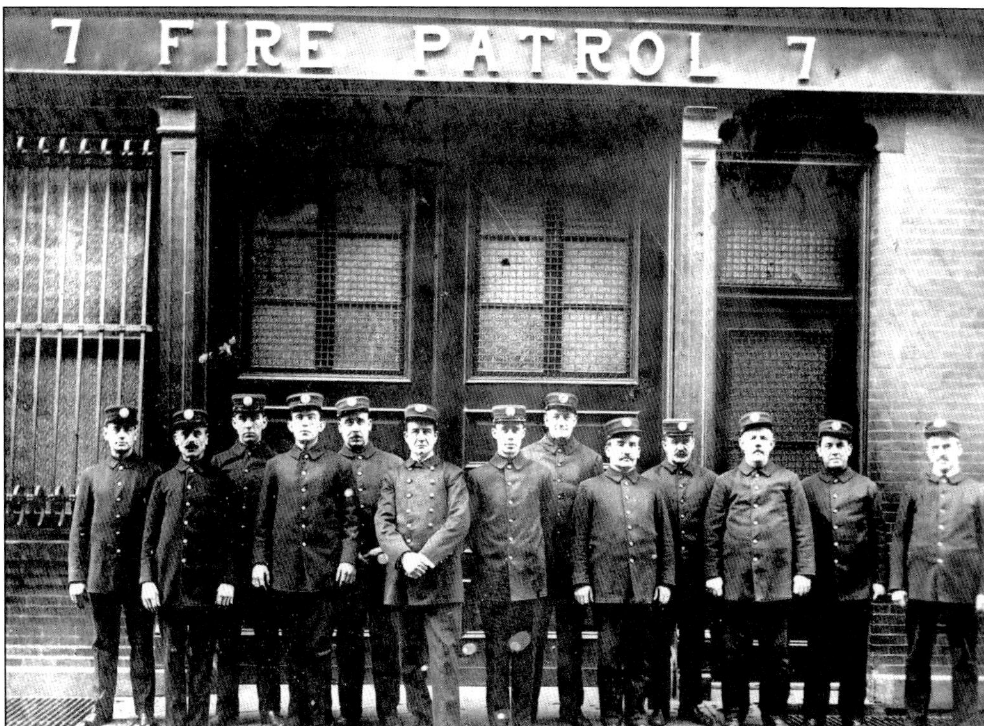

PATROL 7. The Lower Eastside of Manhattan was reported to be one of the most densely populated spots in the country when Patrol 7 was formed in 1906. Patrol 7 was organized on September 3, 1906, at 133 Norfolk Street between Stanton and Rivington. At this time there were over 800 people living on that block alone. Here the company poses for its first known photograph which turned up in the FDNY library archives. The sixth man from the left is none other than Captain Johnston, 40-year veteran of the NYFP. (FDNY.)

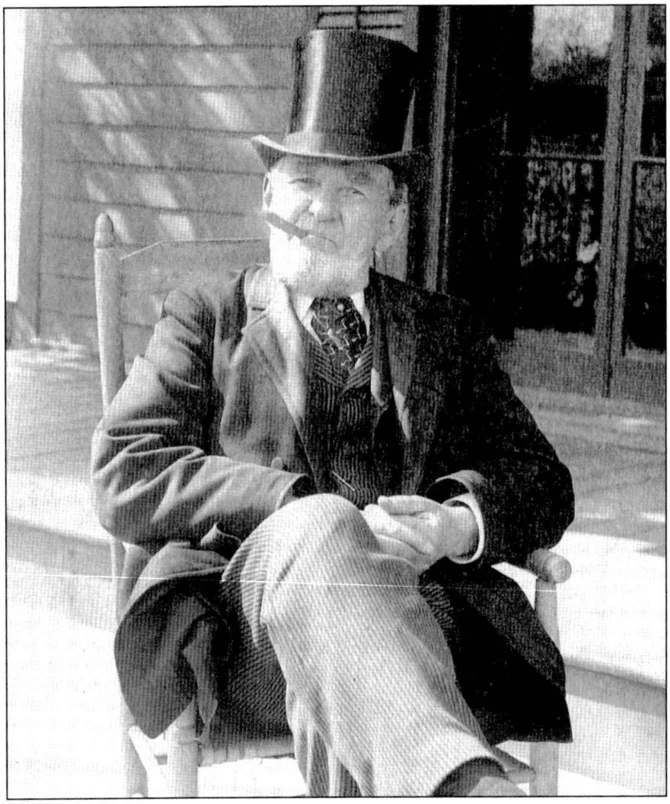

Captain Craig. Capt. John S. Craig joined the insurance patrol by 1839 and was one of the first men employed by the Association of Fire Insurance Companies. He was promoted to captain by 1848, and in 1853 he retired from the patrol. At the time of his retirement he was presented with a silver pitcher and platter set which is still in the possession of his family. (LW.)

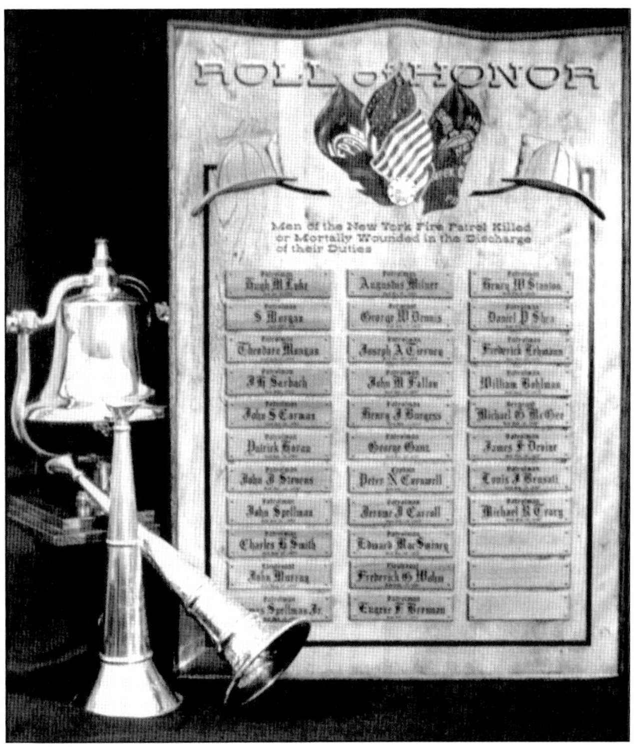

Roll of Honor. During the 1930s, this permanent plaque was dedicated to honor all members of the fire patrol who were killed in the line of duty. Recently two sad additions were made to the plaque: patrolman Daniel Quirk, who died in 1963, and patrolman Keith Roma, who was killed in the September 11, 2001, attacks.

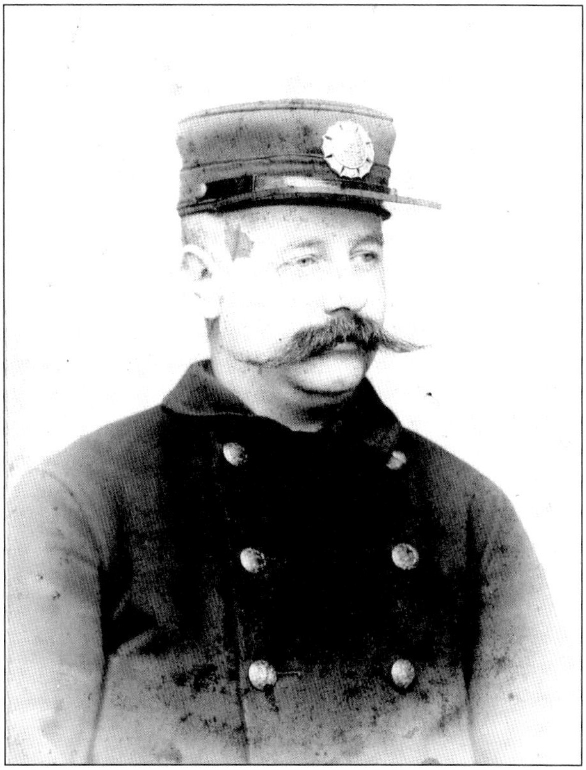

PATROLMAN, 19TH CENTURY. Over the past 170 years members of the NYFP have been known by many names, some of them overlapping as one generation continued to refer to them by the name from the last. Originally known as "the watch," they were also called watchman, regular men, chancemen, red hats, or just patrolmen. This insurance patrolman is wearing the regulation uniform which had been introduced in 1882. (JK.)

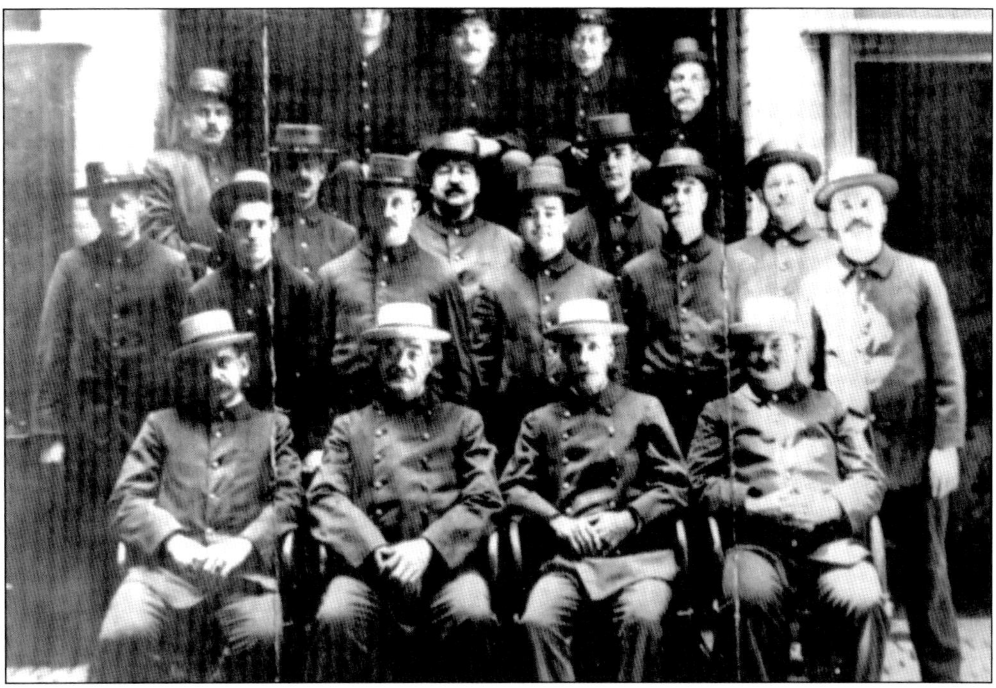

UNKNOWN PATROL, C. 1880. This group of patrolmen and officers are gathered on front of an unknown firehouse in the latter half of the 19th century. It is perhaps the old station of Patrol 3 on West 13th Street which was vacated and sold in 1895. (NYFP.)

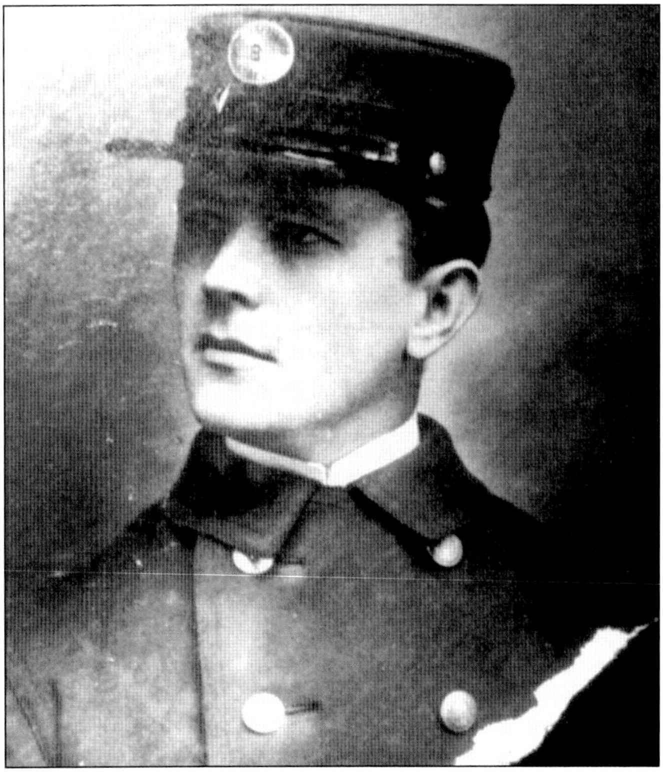

LOUIS HAAG. Patrolman Louis Haag joined the Brooklyn Fire Insurance Salvage Corps (BFISC) on August 1, 1903. He was awarded the Lifesaving Award twice over during 1911 and 1912. He will always be remembered for his courage and generosity. He passed away while handing out ice cream to disabled children at the Holy Name Society picnic in 1933.

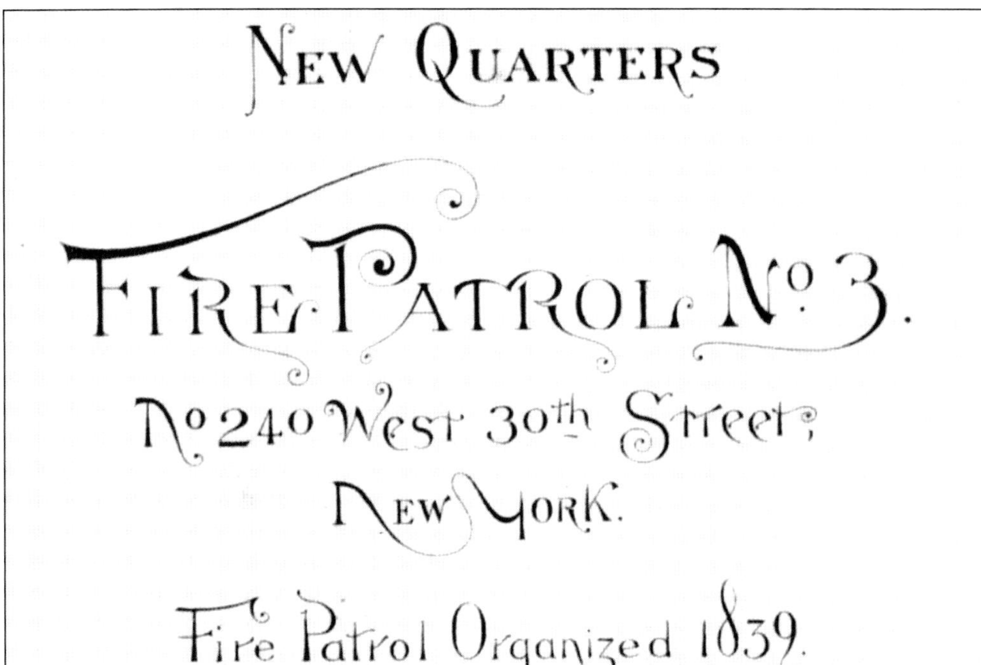

DEDICATION OF PATROL 3, 1895. This card was sent to the many guests upon the opening of the new Patrol 3 in 1895. The building was described as the most beautiful firehouse in the city. Constructed in the Flemish design, it still serves as an active firehouse for Fire Patrol 1.

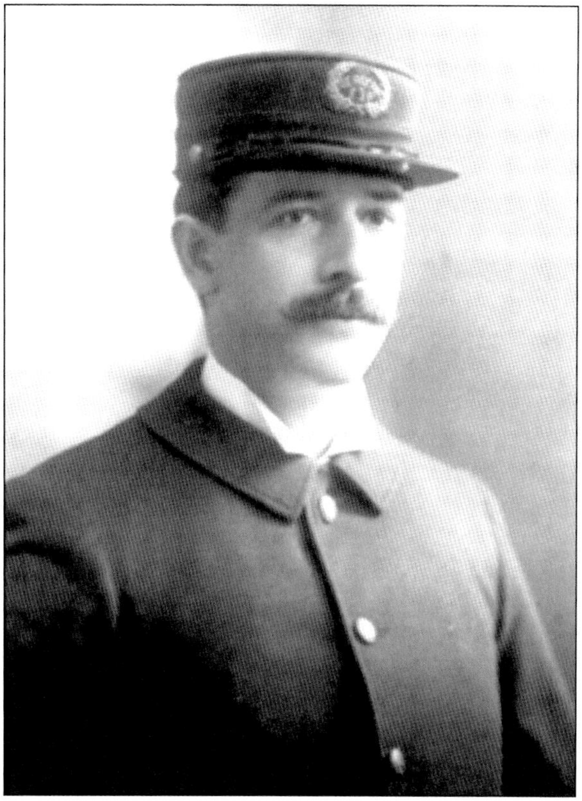

SGT. PETER MAINE, C. 1899. Sergeant Maine was assigned to several different companies in Manhattan during the late 19th and early 20th centuries. He was instrumental in saving several people from fires, however he was never awarded the Medal of Honor. He was, however, placed on the Roll of Merit and continued to serve the patrol until 1919.

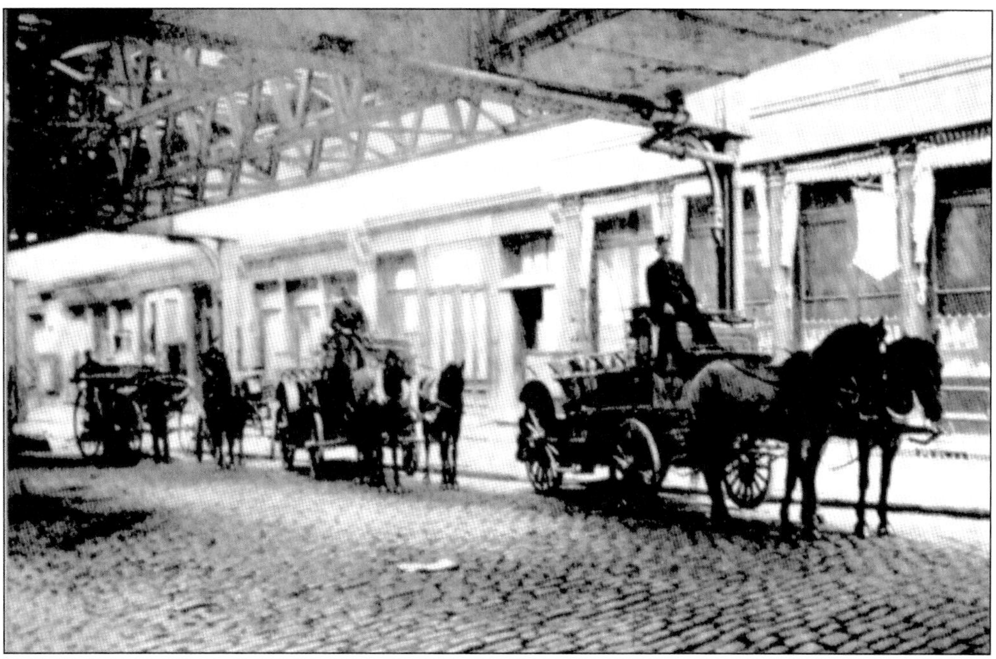

MARY ANN AND PATROL 1, C. 1888. This rare view shows the steam engine of the patrol, and two patrol wagons in front of the old quarters of Patrol 1 at 41 Murray Street. This is one of only two photographs of the *Mary Ann* known to exist, the other is in chapter four.

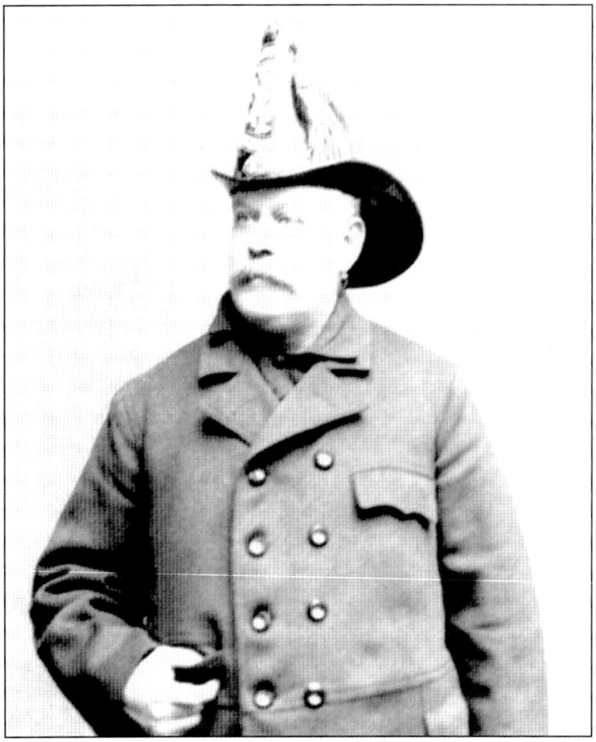

SUPT. ABRAM S. HULL. Abram S. Hull joined the patrol as a captain in 1870. Prior to this time he had served as the foreman of Hook and Ladder Company 6. On March 26, 1886, he was appointed superintendent of the NYFP and served as the commanding officer until April 18, 1899, at which time he was replaced by Frederick Groves. (NYFP.)

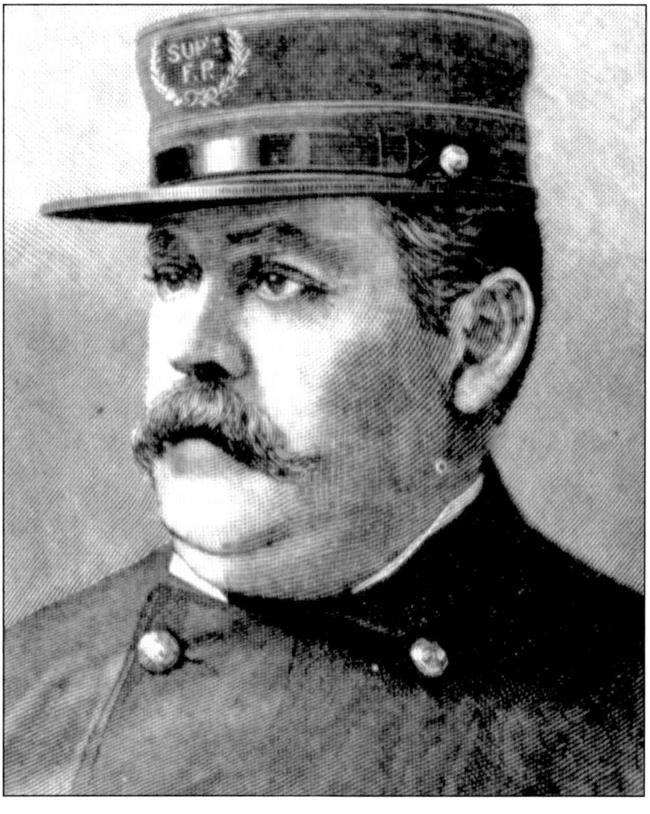

SUPT. ABRAM HULL. Abram S. Hull served as the commanding officer of the NYFP for 13 years. During this time all patrol wagons were redesigned, and he was a silent overseer of the creation of the Brooklyn Fire Insurance Salvage Corps. Abram Hull was born in Manhattan in 1835 and spent nearly his entire life in the fire service.

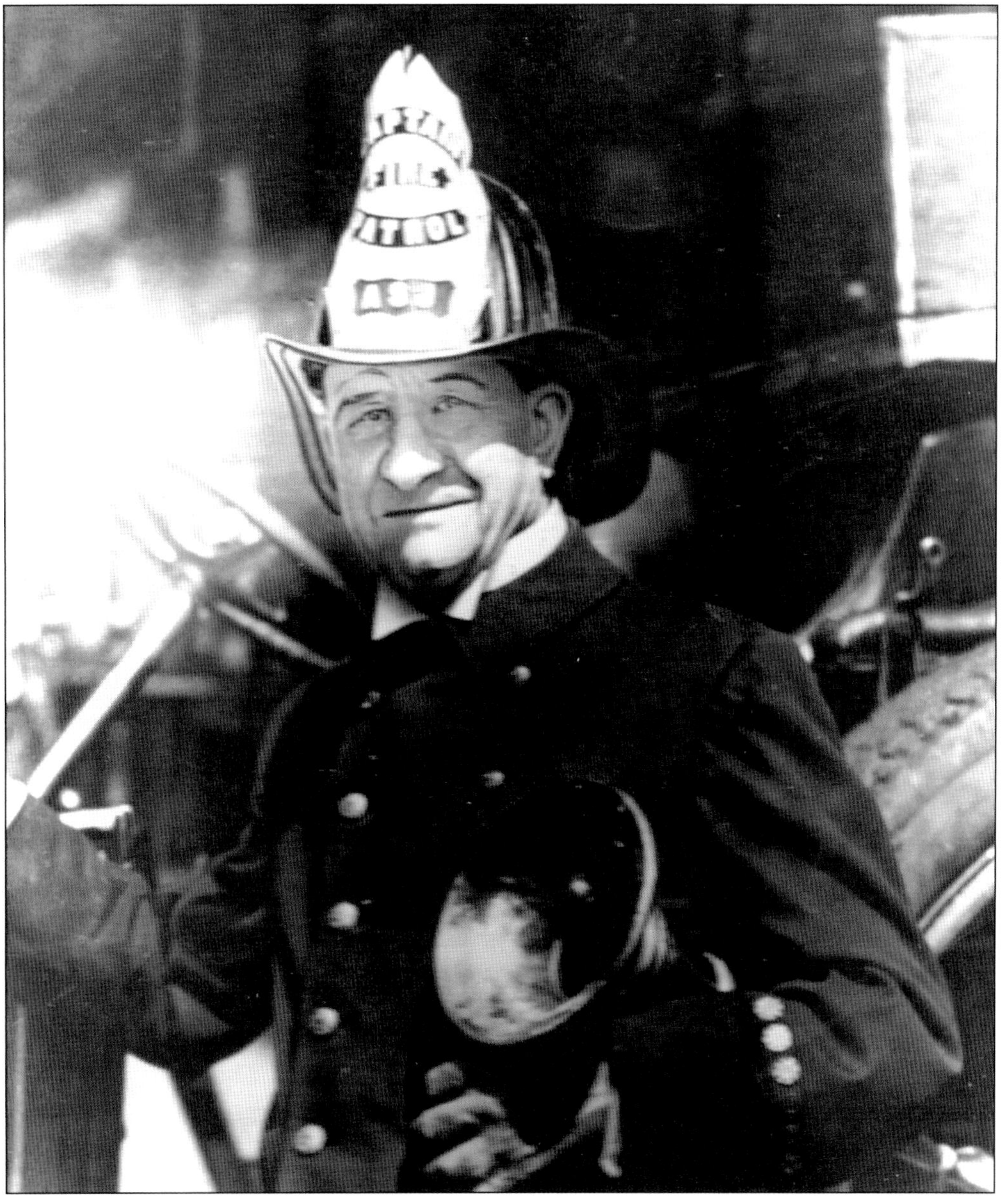

CAPT. A. S. JOHNSTON, 1921. One of the most well known fire patrolman and firefighters in the history of New York City was Albert Sidney "Sid" Johnston. Captain Johnston joined the patrol around 1887. The son of an original member of the fire police of 1835, Johnston was a man of vast fire service knowledge and was well respected by all. He was often placed in service as an acting battalion chief at fires and won the FDNY lifesaving medal on three separate occasions, one of them, two years before his retirement in 1925. In 1912, Johnston arrived at the Equitable Building fire and quickly made his way to the fourth floor. Upon arrival, he gave some quick advice to Chief William Walsh who was working the fire, "Better get your men out of here Chief," yelled Sid. The chief yelled back "for what, we've got this thing licked," three weeks later they finally found Chief Walsh's body in the rubble. Today, the Albert S. Johnston Medal is awarded by the FDNY for acts of extreme lifesaving to either, firefighters, or fire patrolman, a fitting tribute to a brave man who served 40 years with the NYFP. (NYFP.)

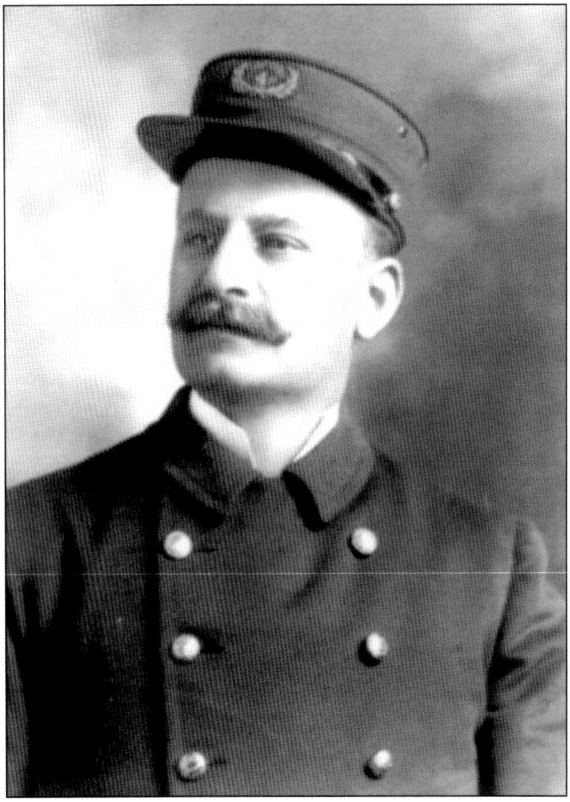

UNKNOWN PATROLMAN, C. 1899. This officer's name has been lost to obscurity. However by his uniform we can tell that he served in the time period from 1882 to 1922. In the early 1920s the high button coat was changed, and a tie was added to the uniform code.

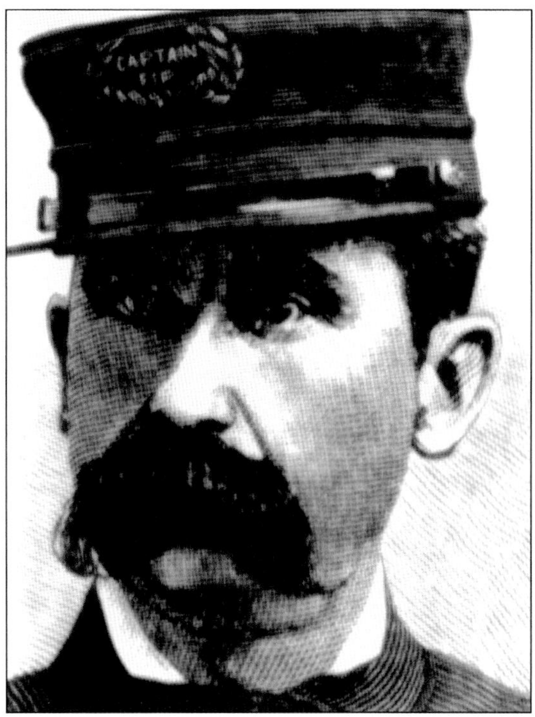

FREDERICK GROVES. Frederick Groves joined the insurance patrol during the early 1870s. He was first assigned to Patrol 1 on Murray Street, became a sergeant in 1883, and was later promoted to captain. By 1911, Groves had been promoted to superintendent and was charged with the supervision of six fire patrol units in Manhattan and three in Brooklyn. (LW.)

UNKNOWN PATROLMAN, C. 1928. Standing outside the Dean Street firehouse of Patrol 8 is this dapper fellow who may have been on housewatch duty that day. Patrol 8 eventually became Patrol 3 in 1955 after the closing of Patrols 1, 5, 6, 7, and 10. It remains today. (NYFP.)

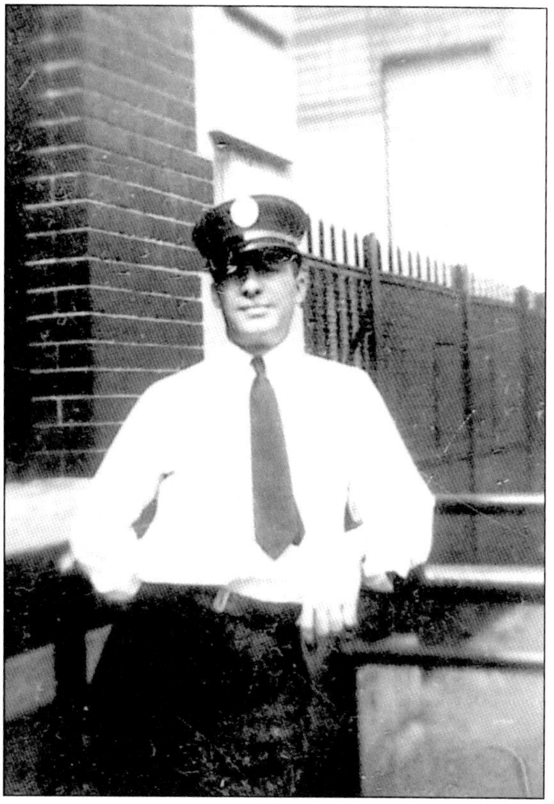

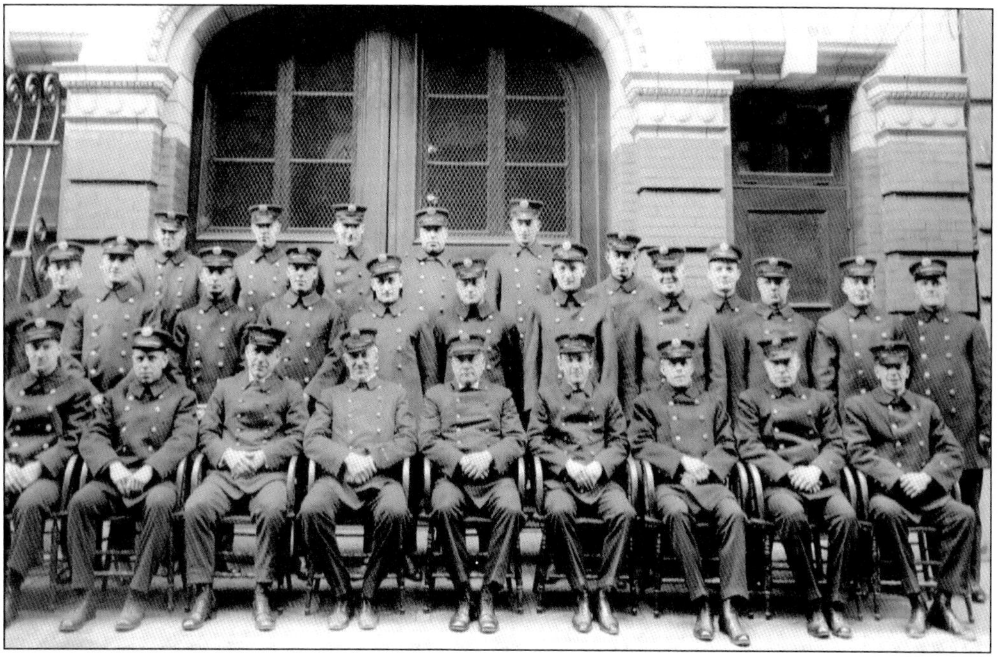

FIRE PATROL 2, 1918. Over 30 members from both sections of Patrol 2 are assembled here in front of the West 3rd Street House. Fire Patrol 2 has a long standing tradition of posing for company portraits.

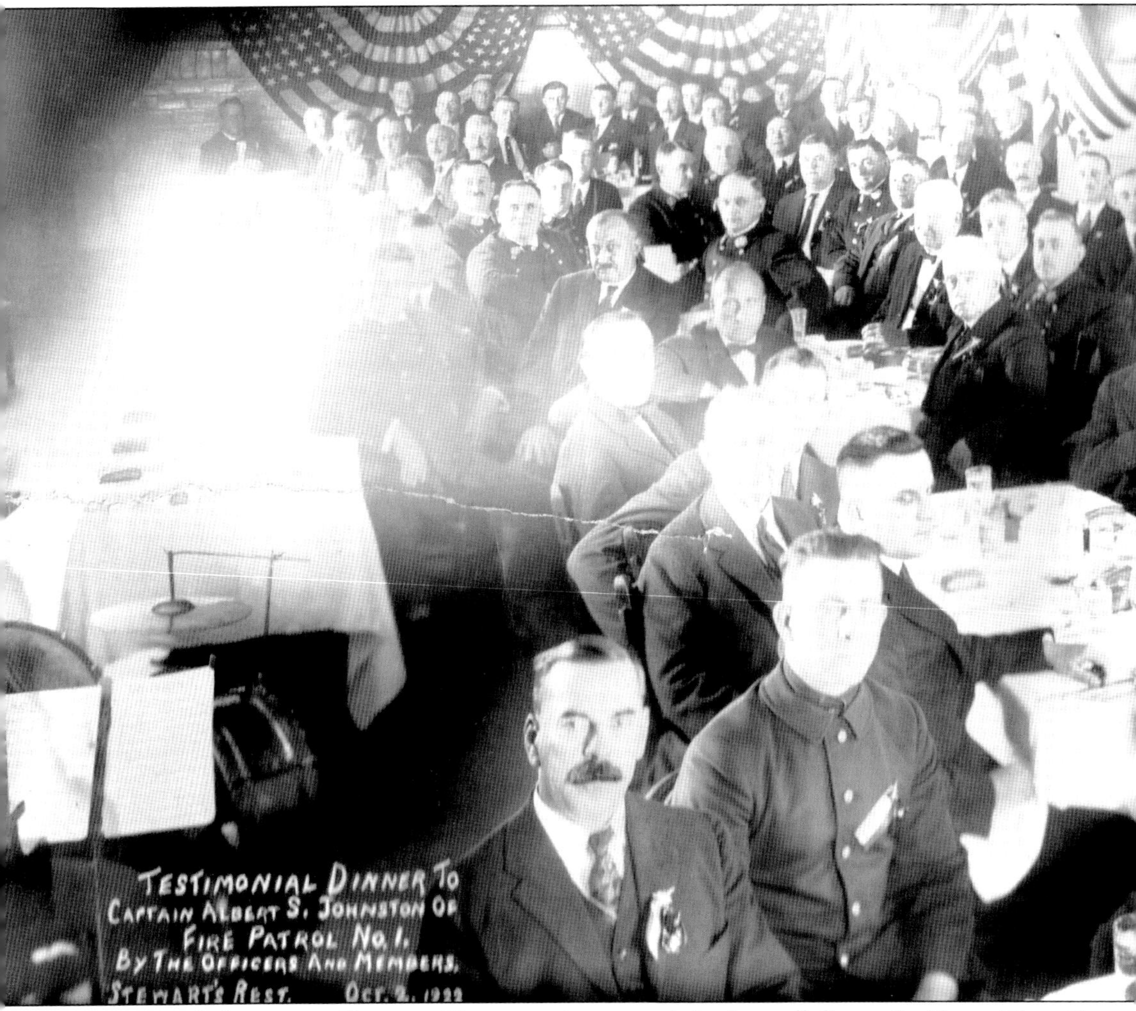

ALBERT S. JOHNSTON DINNER. This portrait captured the farewell dinner for Capt. Albert S. Johnston on October 2, 1922. The event was held at Stewart's Restaurant in Manhattan. The portrait was presented to the fire patrol by the family of Sgt. Christopher F. Young.

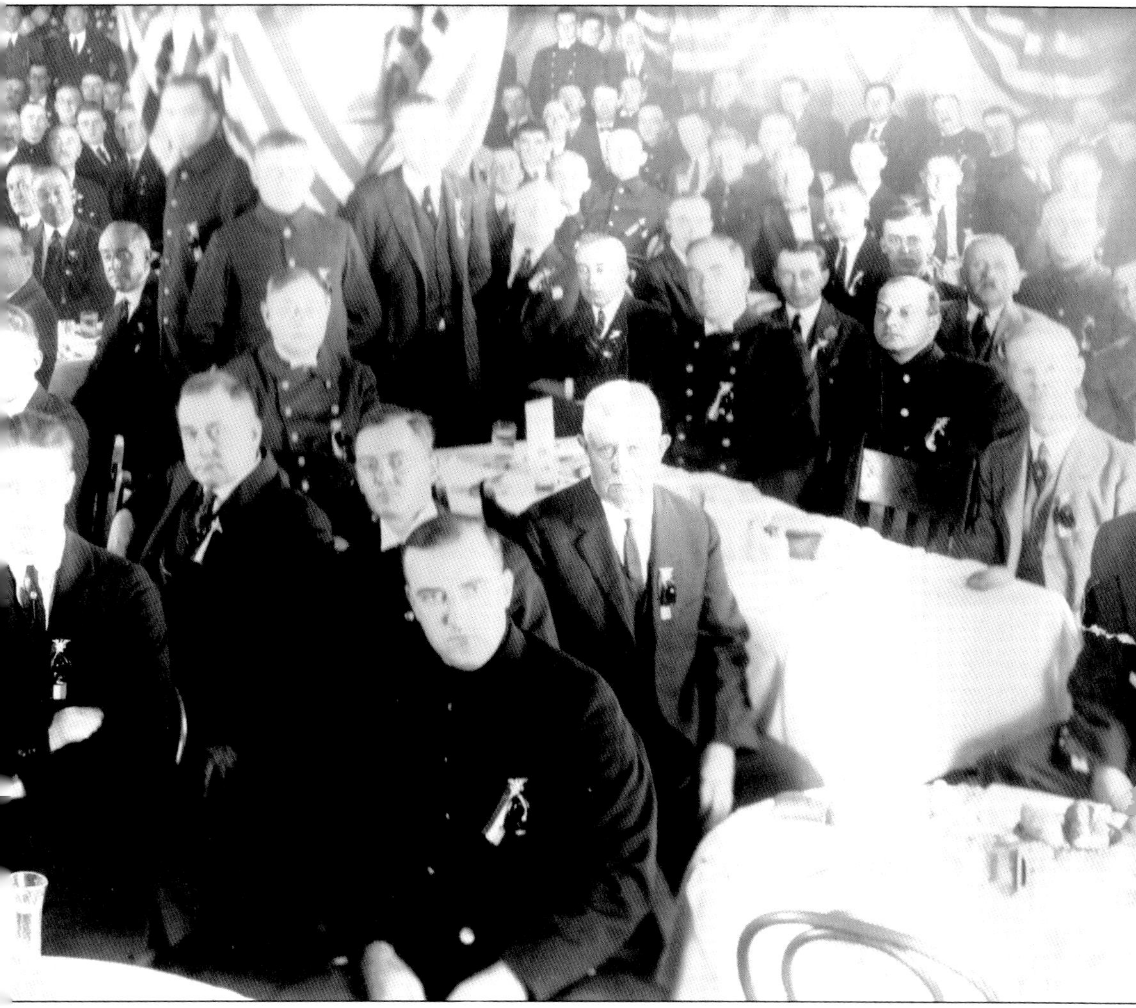

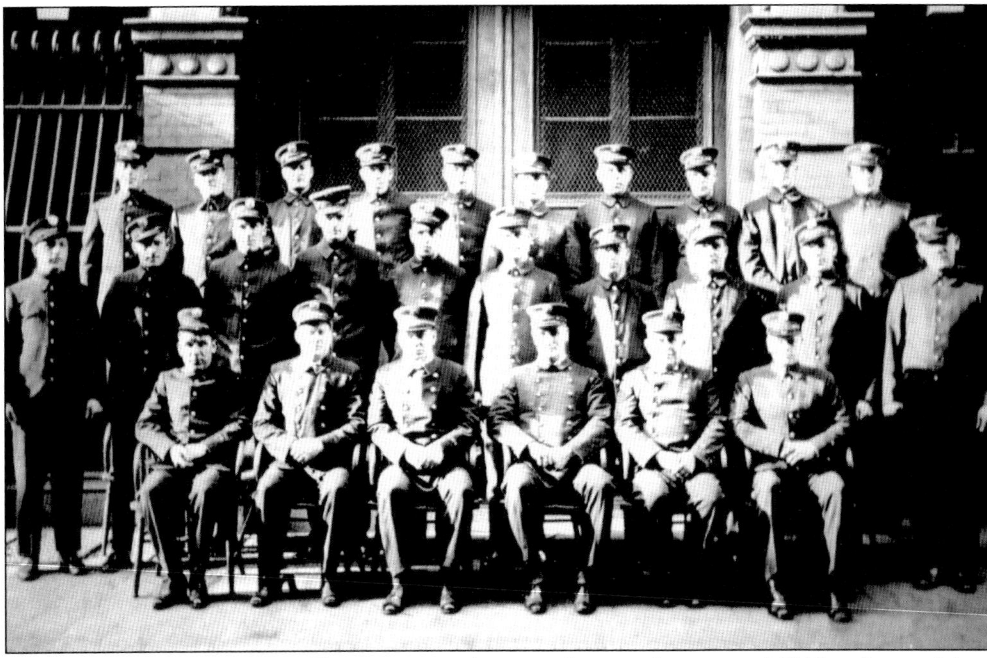

GREENWICH VILLAGE MEN. This company photograph was taken around 1913 in front of their quarters. Patrolmen at this time were working 10 days on, followed by 1 day off. Of course members wishing to have dinner were allowed a 3 hour meal break each day. (JK.)

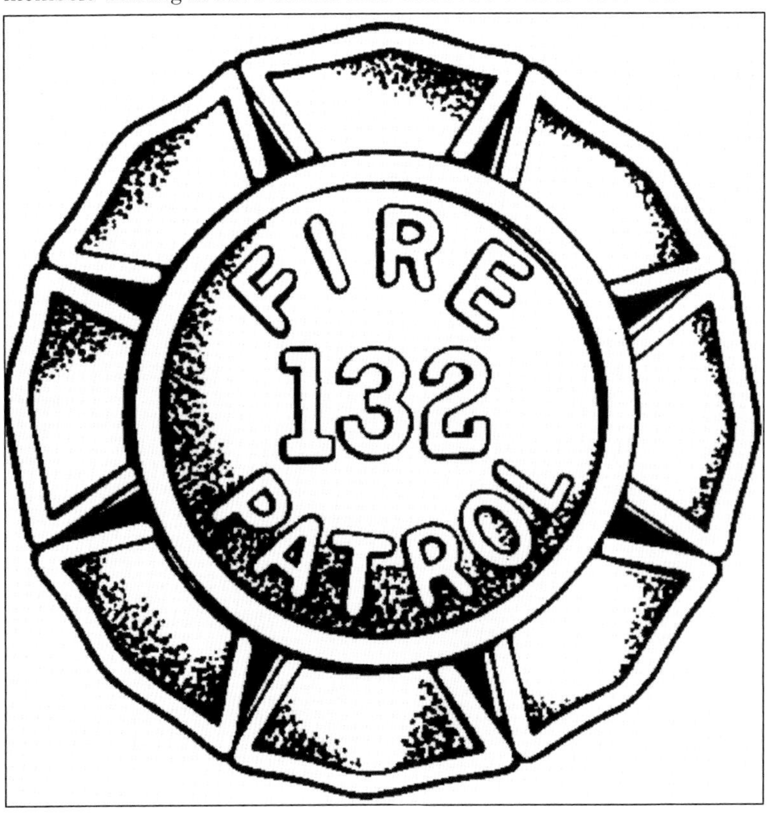

BADGE OF THE FIRE PATROL. For the first time in the history of the patrol, a former deputy chief of the FDNY was hired as chief of the fire patrol. In 1952, Joseph Scanlon was appointed chief and one of his first projects was to adopt the double Maltese cross as the new regulation breast badge. However, the double maltese had been is use as the hat badge for many years.

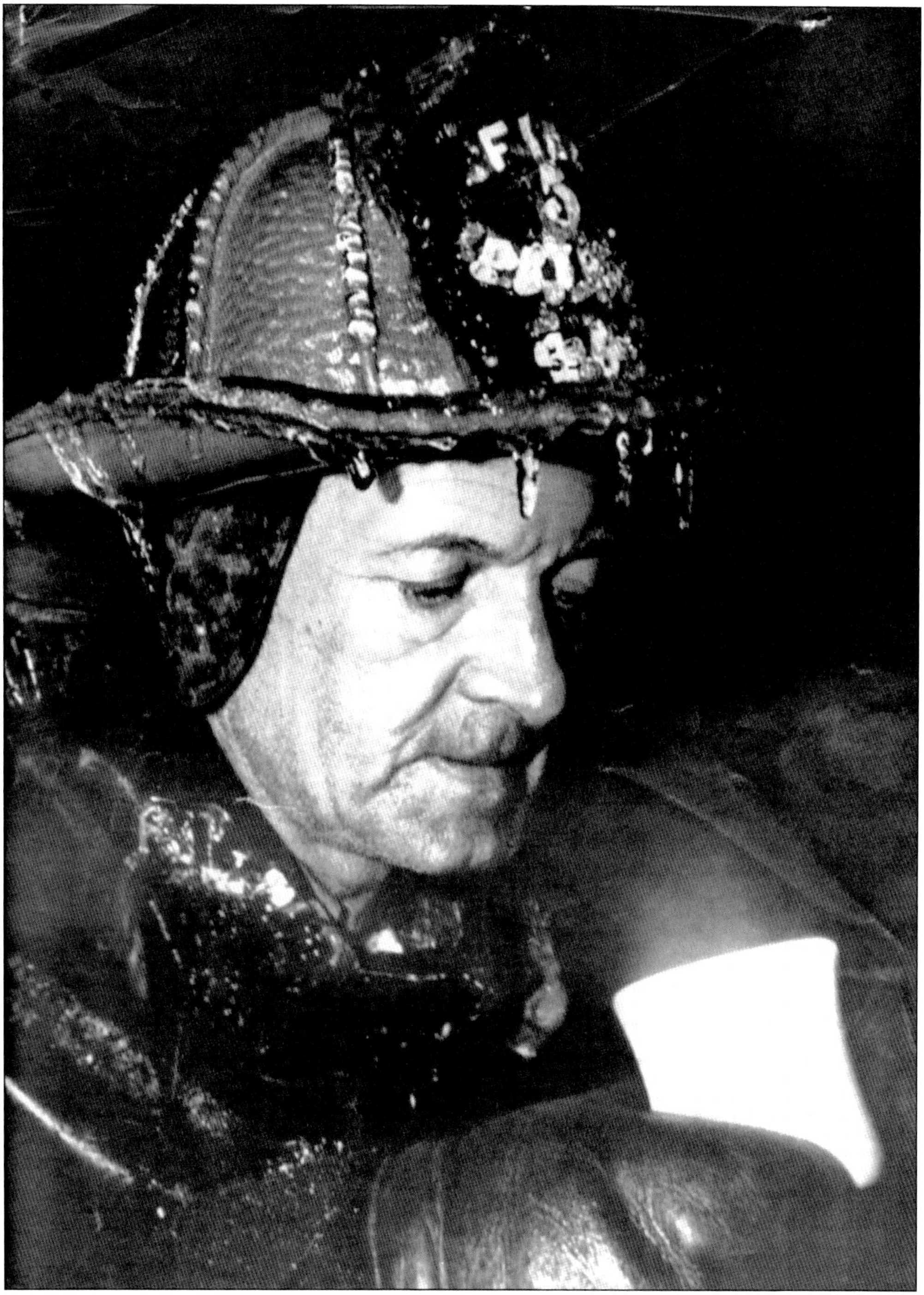

Braving the Elements, 1955. Hot coffee is a welcome site to any patrolman in the winter months. Cold winter days always bring long busy nights for the patrol. In the freeze of December 31 through January 3, 1962, the patrol responded to hundreds of broken pipes, fires, and emergencies. All total the patrol processed over 1,300 alarms in four days. It had been the busiest time for the job in over 30 years. (LM.)

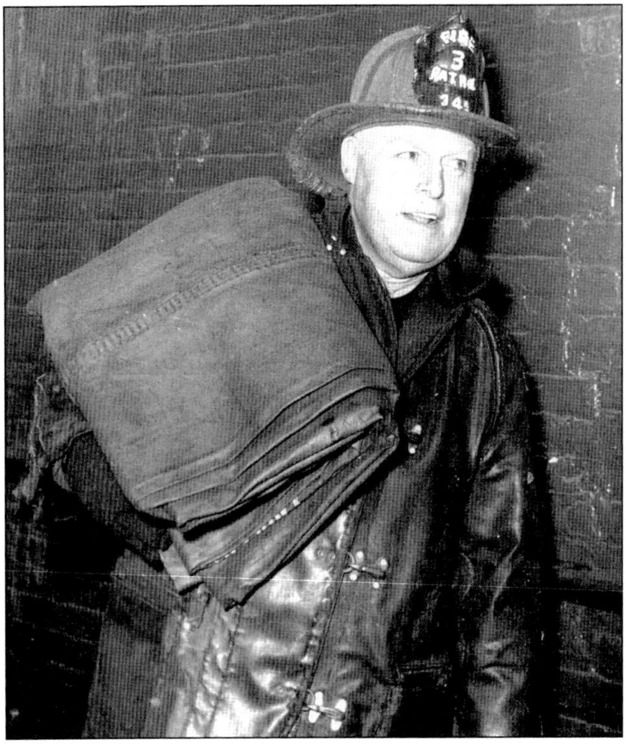

PATROLMAN, 1949. A member of Patrol 3 heads into another "job" with two covers. His rubber coat and red helmet were ready to protect him in the days before self contained breathing apparatus. At every fire, patrolmen shoulder two covers, carry a set of "irons" (axe and halligan bar), lights, and other equipment as the scene dictates. (FDNY.)

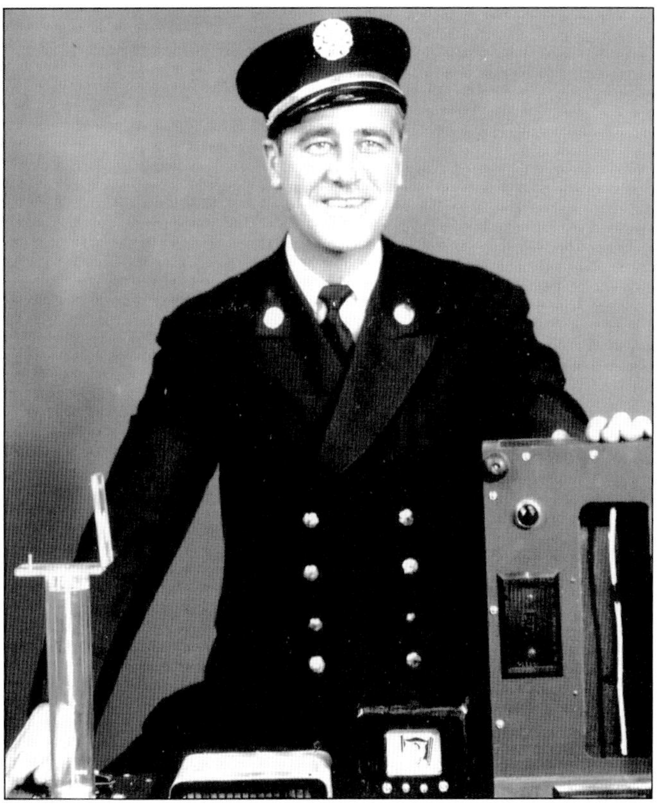

SGT. JOE LAVERY, 1949. Fire prevention programs for the cities schoolchildren have prevented thousands of untold fire injuries and deaths since the program's inception in 1920. Here, another demonstration is being photographed for citywide public relations in the 1950s. (NYFP.)

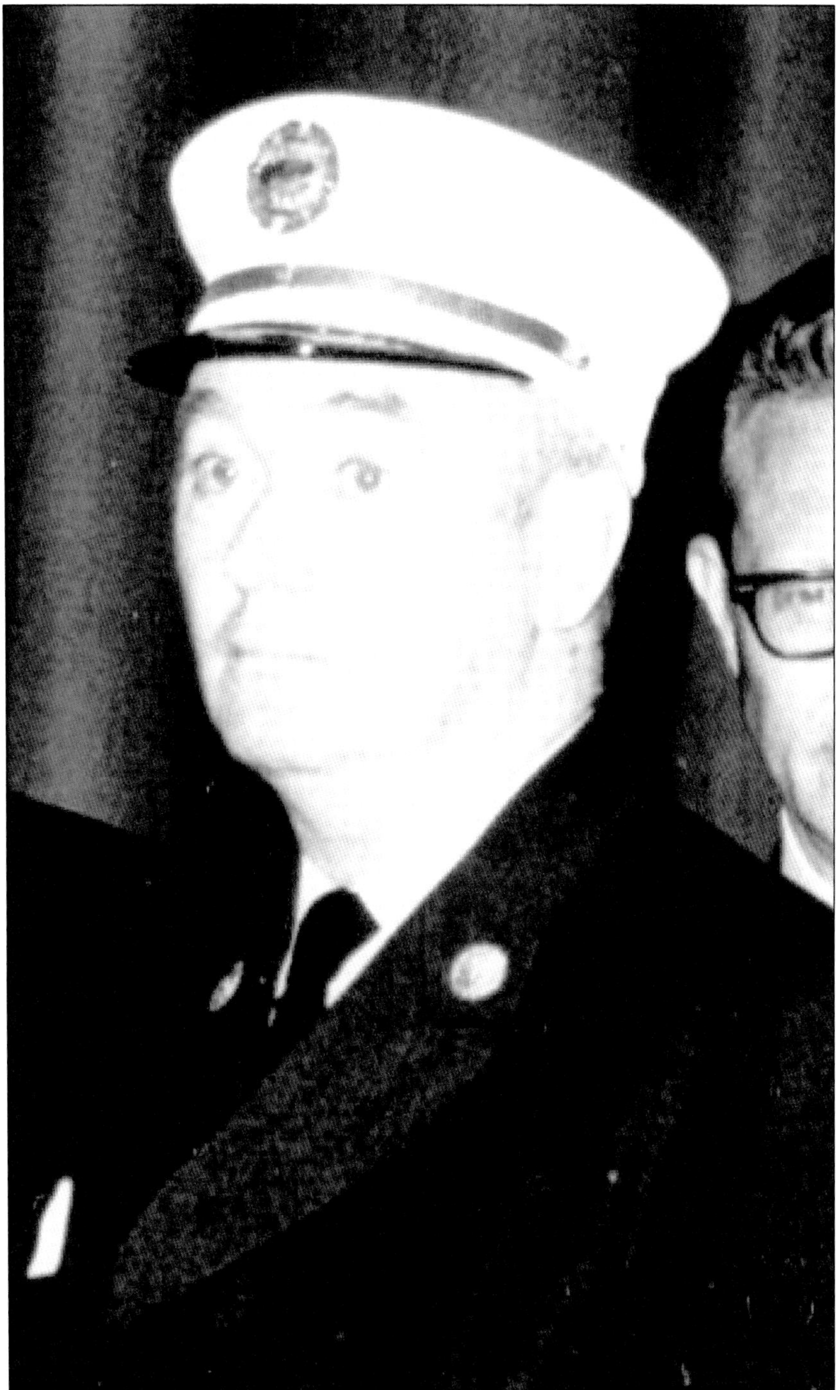

CHIEF LYNCH. James P. Lynch joined the NYFP on June 16, 1948, and served through all ranks, including captain, until 1967. In February of that year, James Lynch was appointed as the chief of the fire patrol, a position which he proudly held until his death on October 8, 1984. At which time, Francis Dempsey assumed the position of chief. (NYFP.)

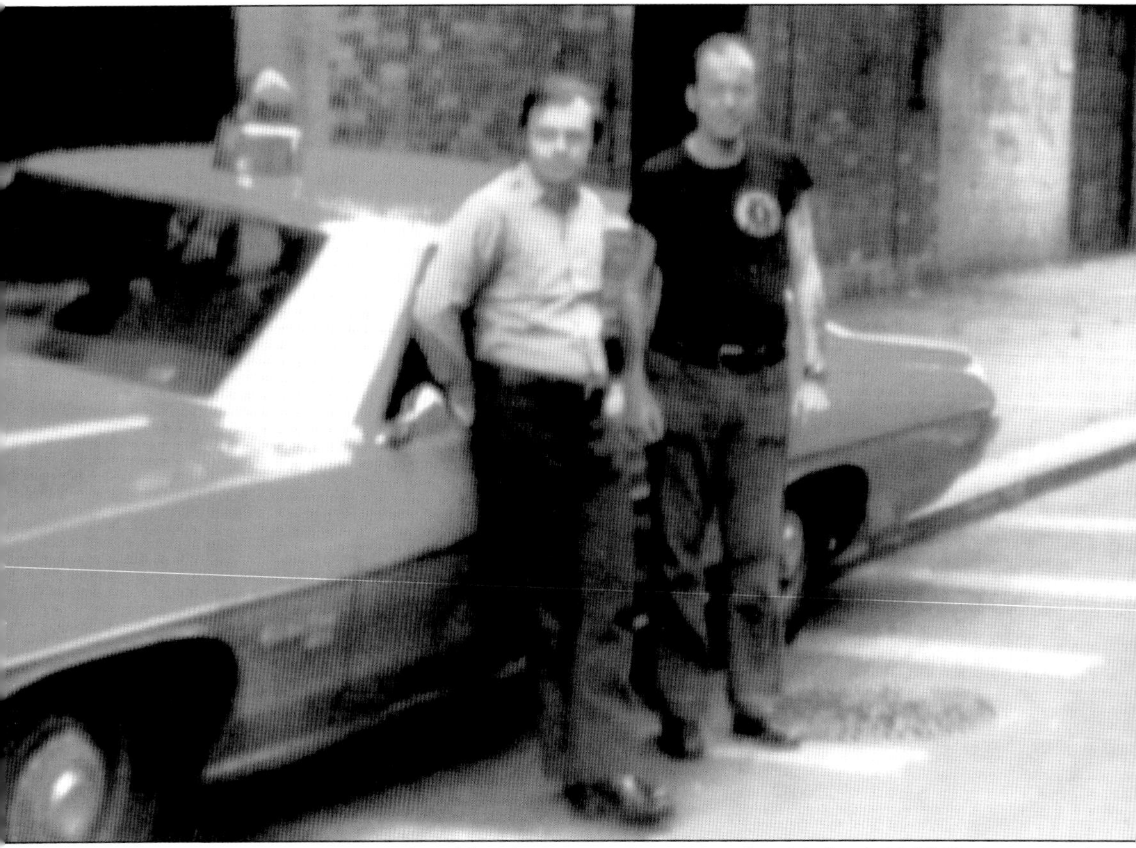

SUMMER, 1973. Leaning on the old 1969 Pontiac Car 100 is Sgt. Michael "Yak" Yackinicz and patrolman Bill Tobin. Tobin joined the NYFP in 1967 and was a second generation member. Retiring in the late 1990s, Yackinicz joined the patrol on July 14, 1966, and 20 years later, he went on to become a FDNY fire alarm dispatcher. (FL.)

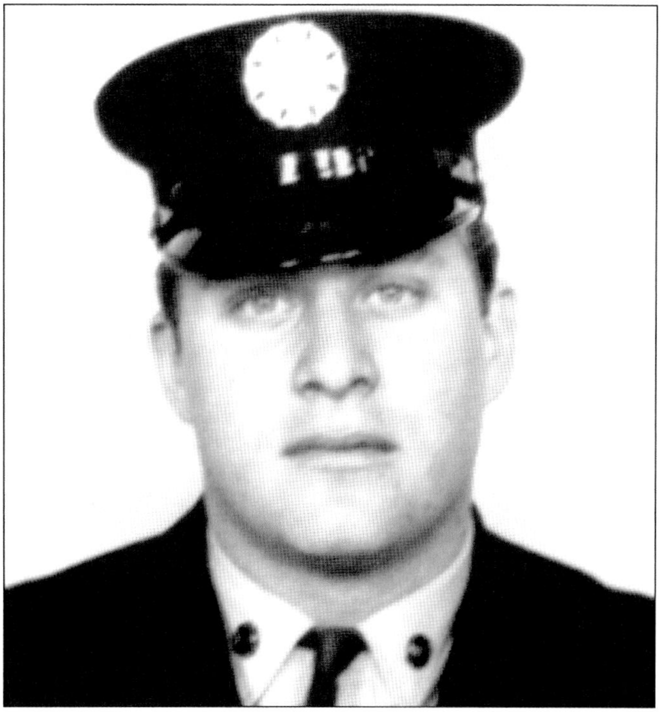

PATROLMAN EDWARD POSPISIL. Edward Pospisil was appointed to the fire patrol on December 16, 1964. Just 18 months later, patrolman Pospisil is seen at Box 598 at 7 East Street (below). During this blaze the floors of the Wonder Drug Store gave way, trapping 10 firefighters and burning two others. During the recovery operation, patrolman Pospisil drew a detailed map of the basement that eventually led to the recovery of the 10 fallen fireman. In 1967, he was presented with the highest honor that the FDNY can bestow on a fire patrolman, the Albert S. Johnston Medal. (FDNY.)

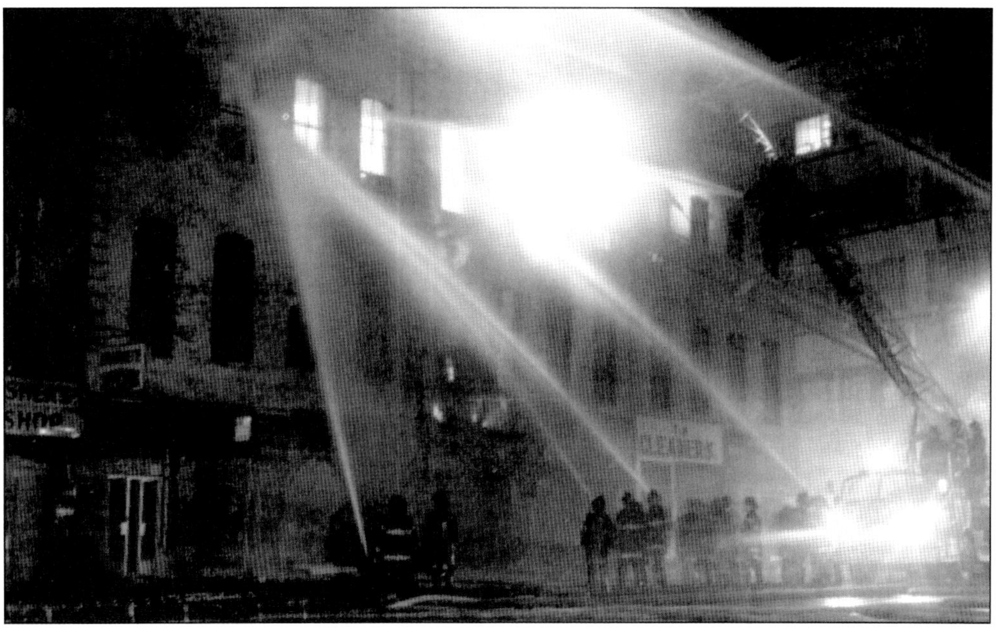

EAST 23RD STREET FIRE. The Wonder Drug Store was located at 6 East 23rd Street and was known as exposure two during the alarm. The original fire building was a brownstone with an art gallery on the first floor. The fire, which had started in the basement of No. 7, had been slowly burning the floor beams of the Wonder Drug Store. At 10:39 p.m., the floor collapsed without warning in the nearly 100-year-old building and plunged 10 firefighters into the fiery basement, two others were killed in the ensuing fireball on the first floor. It was the worst loss of life in the FDNY history until September 11, 2001. (FDNY.)

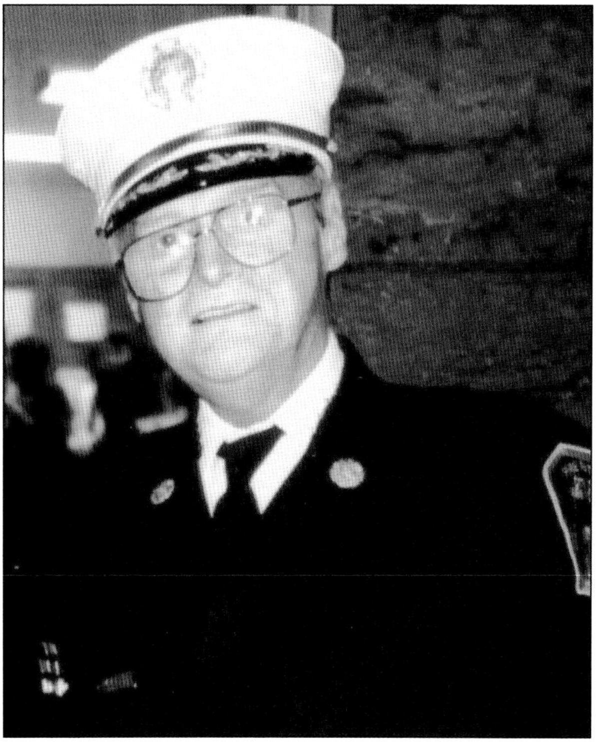

CHIEF FRANCIS DEMPSEY. Chief Dempsey was appointed chief of the patrol in October of 1984, after the death of Chief Lynch. Dempsey had joined the patrol in 1959 and worked his way through all the ranks as well as serving during the riots and blackouts of the 1970s. In 1996, he retired and passed away a few years later. (LW.)

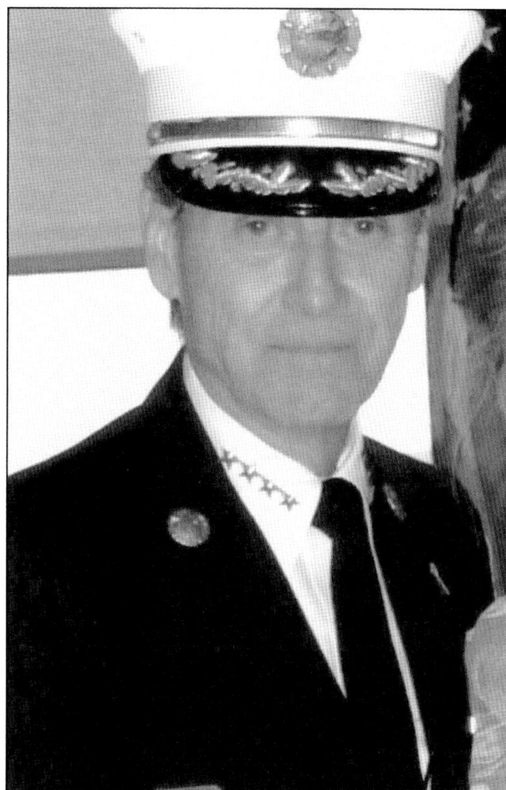

CHIEF RAYMOND CARROLL. Chief Raymond Carroll assumed the duties of fire patrol chief in 1996 after the death of Chief Francis Dempsey. Carroll joined the patrol in 1968 and spent most of his career assigned to Patrol 1 on West 13th Street. After many years as captain, he was promoted to deputy chief in 1988. A survivor of September 11, Chief Carroll said goodbye to the patrol and his beloved men in March 2005, during the patrol's 170th anniversary. He retired to the Bronx where he promised his wife, Cathy, a good vacation. (NYFP.)

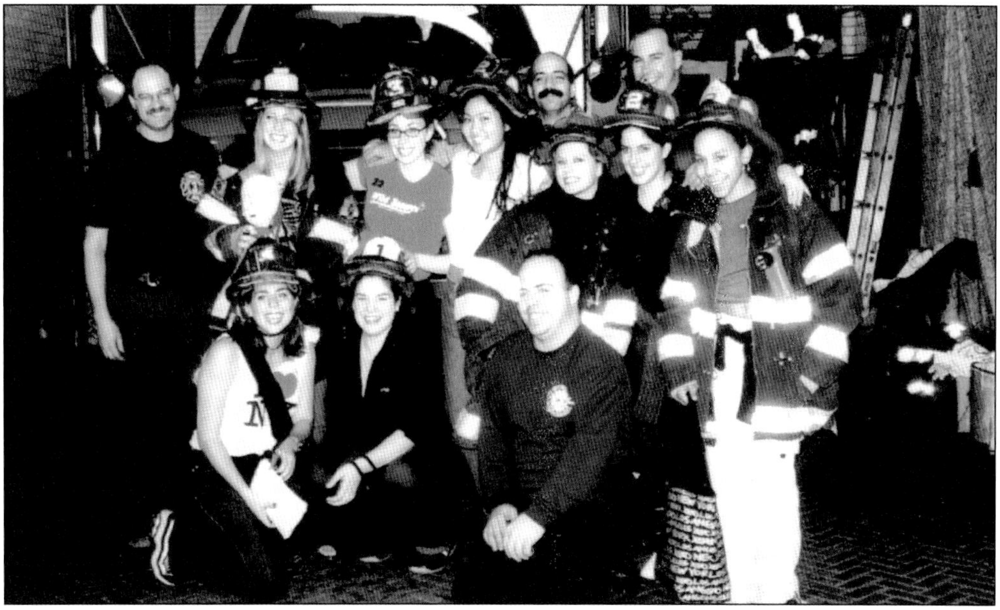

PATROL 2 VISITORS. Many tourists who stop at Fire Patrol 2 honestly believe that it is a prop, or a museum, as Greenwich Village continues to be a remarkable place for tourists to visit. Many are surprised when they learn it is an active firehouse dedicated to protecting the cities property. Here, a group of Midwest college students stop by for a visit. On this day, Sgt. Larry Wagonfeld, and patrolmen Tony Emmanuelle, Eric Schultz, and Pat Mooney posed with the tourists. (TER.)

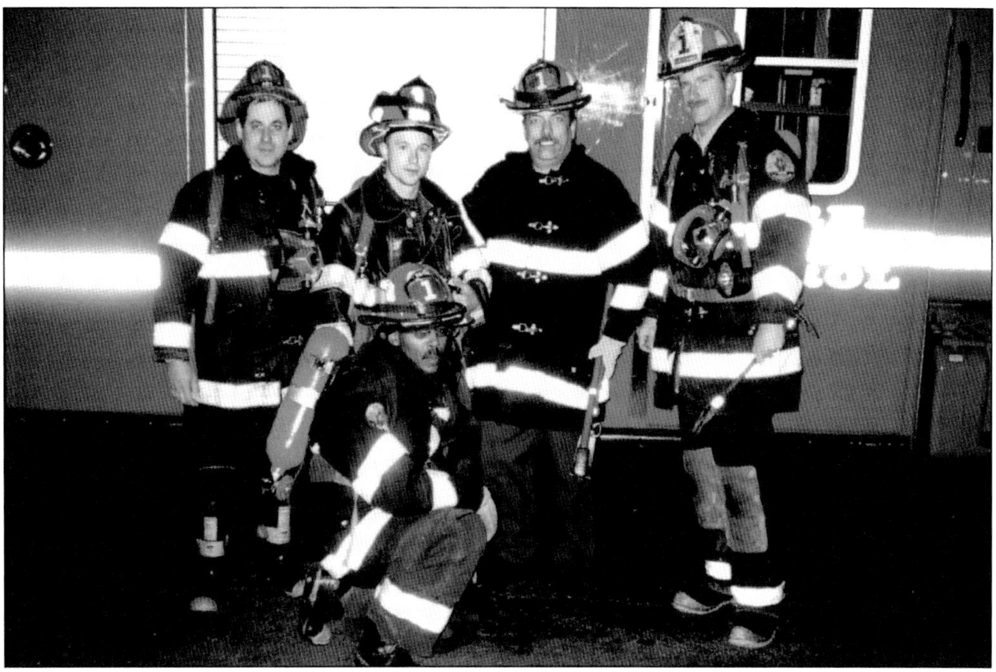

PATROL 1 GROUP. Here, the members of Patrol 1 stop for a quick picture after taking up at a mid-town Manhattan blaze. From left to right are patrolmen Mike Markey, John Pernie, Bobbie Joe Franks, Lt. Tom McGee, and Mike Horton (kneeling). (TM.)

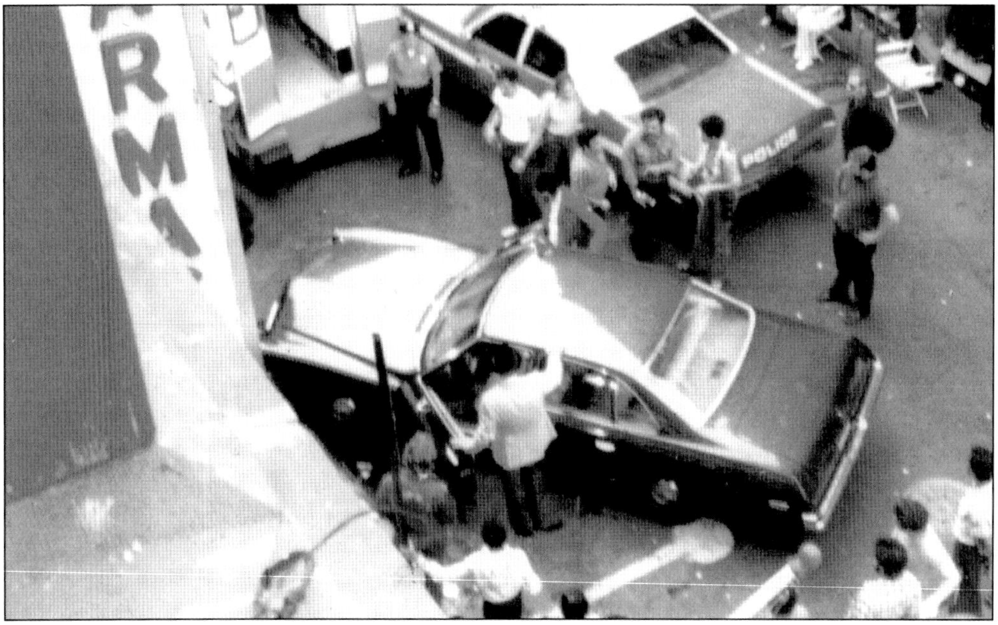

KOJAK AT THE PATROL. From 1973 to 1978, Telly Savalas portrayed the tough talking New York City detective, Kojak. In 1974, the set of *Kojak* moved to West 3rd Street to be filmed directly in front of the quarters of Patrol 2. This photograph, taken from the third floor of the firehouse, shows Savalas exiting his car for another "Mob" investigation in the streets of Greenwich Village. (FL.)

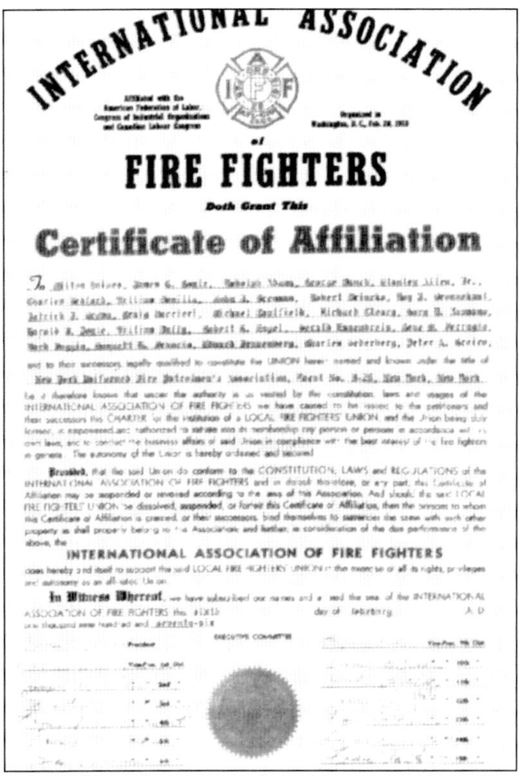

INTERNATIONAL ASSOCIATION

of

FIRE FIGHTERS

Doth Grant This

Certificate of Affiliation

INTERNATIONAL ASSOCIATION OF FIRE FIGHTERS

In Witness Whereof

IAFF CHARTER. The patrol has been affiliated with a union since the early 1940s. It was not until 1976 that the membership was awarded a chapter of the International Association of Firefighters. In that year, Local I-26, also known as the UFPA (Uniformed Fire Patrolman's Association) was formed with 26 members signing the charter. In the following years, the IAFF has remained as the patrol's bargaining unit and has helped the patrol stay current with salaries and wages. Only two of the charters original signers remain, Capt. Pete Greico and Sgt. Milton Snipes. (UFPA.)

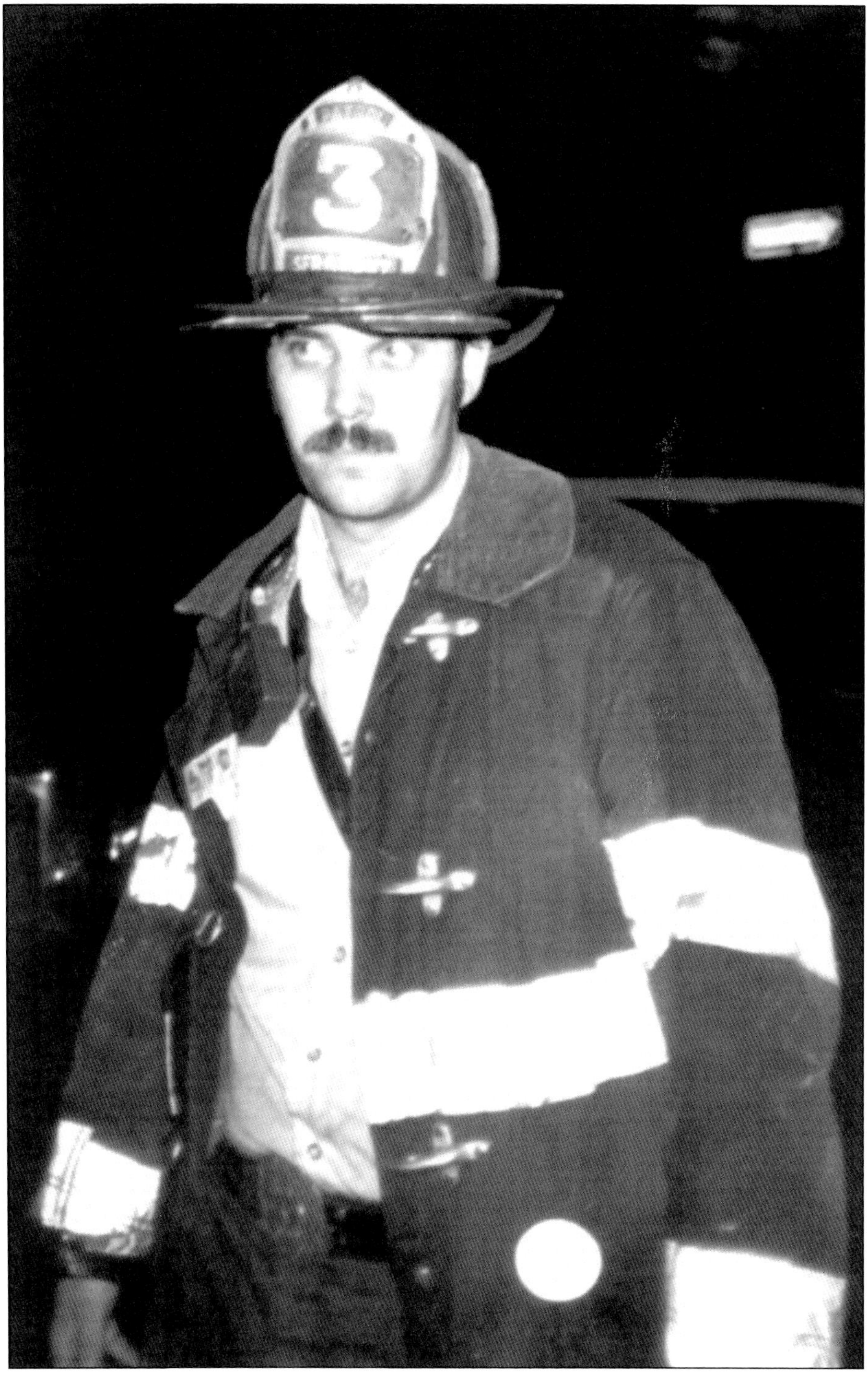

SGT. RICHARD HEFFERNAN. Rich Heffernan was appointed to the NYFP on January 3, 1973, after serving in the United States Army. He was assigned to Patrol 1 for many years and was appointed deputy chief in 2002. Upon the retirement of Chief Raymond Carroll in 2005, Richard Heffernan proudly assumed the duties as chief of the New York Fire Patrol. (RH.)

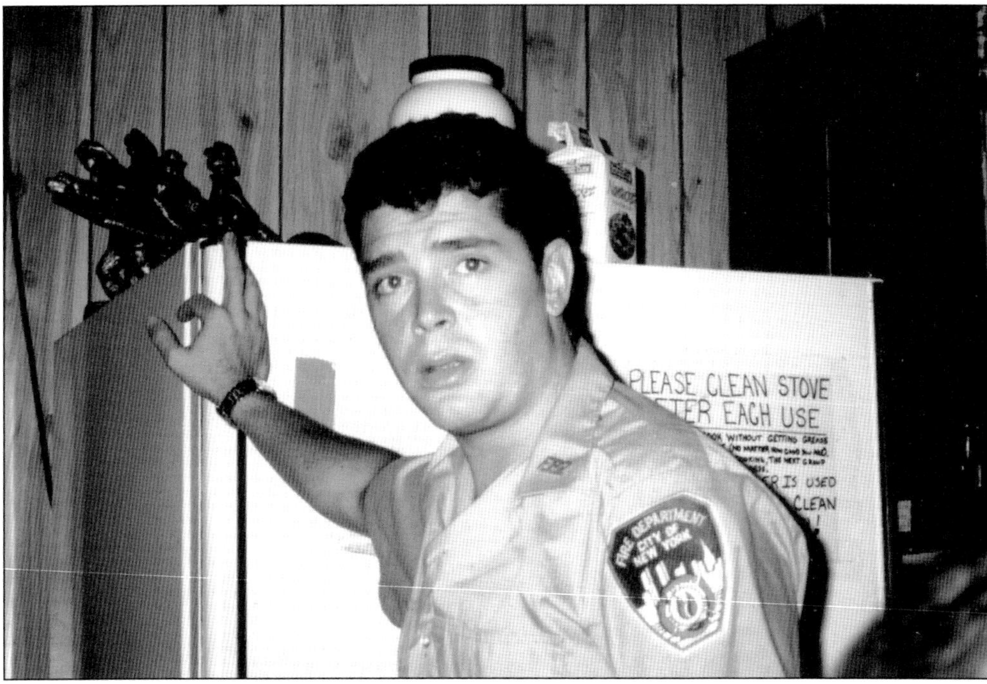

Joe Crawley. Joe Crawley was assigned to Fire Patrol 2 in Greenwich Village. In later years he would serve as a chief's aide, and was promoted to sergeant in 2004.

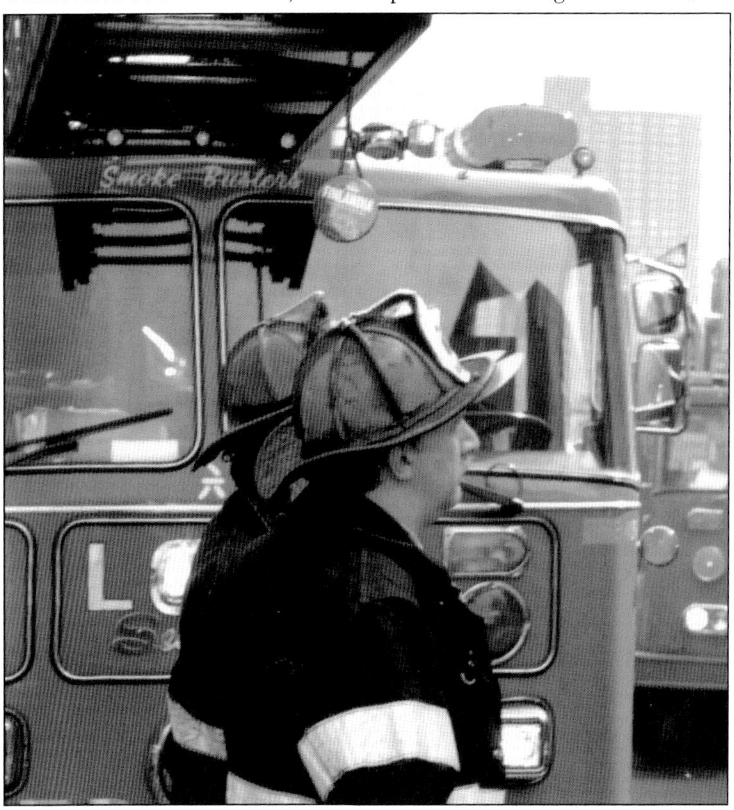

Watching and Waiting. Many times there is no possible way of undertaking salvage operations. Often fires that are of incendiary nature burn extremely hot and fast and will destroy all contents before the patrols arrival. At these fires the men will then go to work in the adjoining buildings that are threatened by fire and water damage. (FL.)

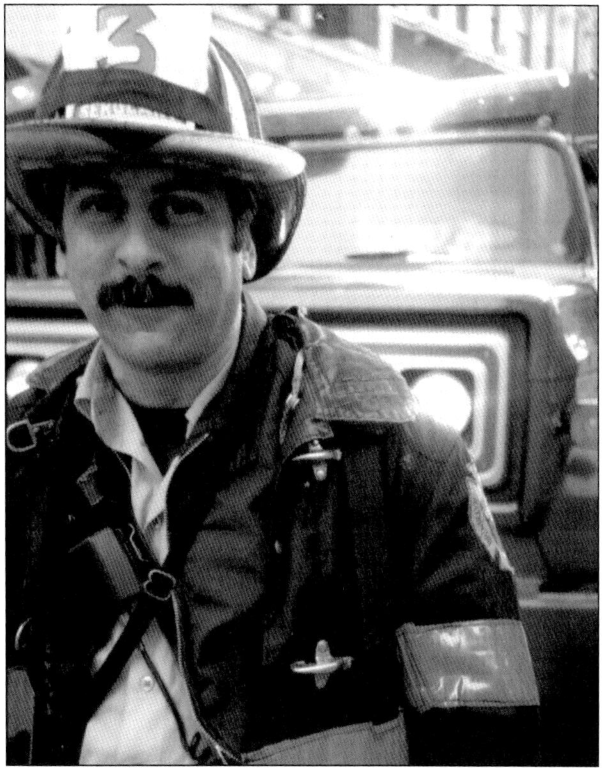

SGT. DENNIS KENNY. Dennis Kenny joined the NYFP on August 20, 1969, after serving several tours as a decorated Marine during the Vietnam War. At one time he left the patrol for a short stint with the FDNY dispatchers. He returned to the patrol and was promoted to the rank of sergeant. (DK.)

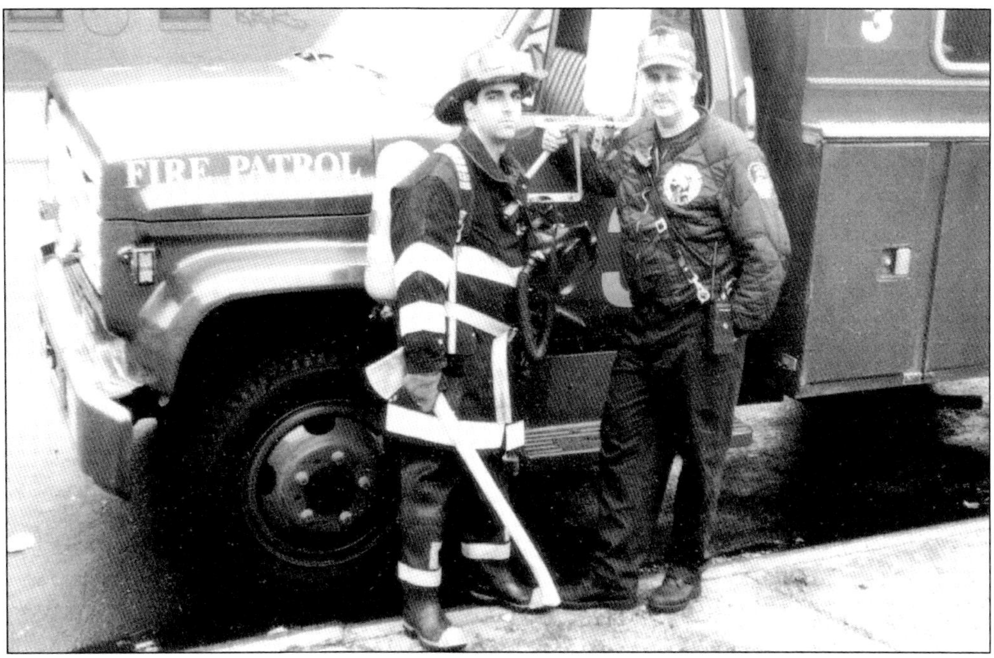

PATROLMEN MARGARITIS AND CURRAN. This picture was taken after a smoky Brooklyn fire in 1993. On the right is patrolman Paul Curran "best wheel man on the job" and patrolman Pete Margaritis who served with the patrol for several years. They are shown in front of Car 48, a 1986 Chevy/Stahl.

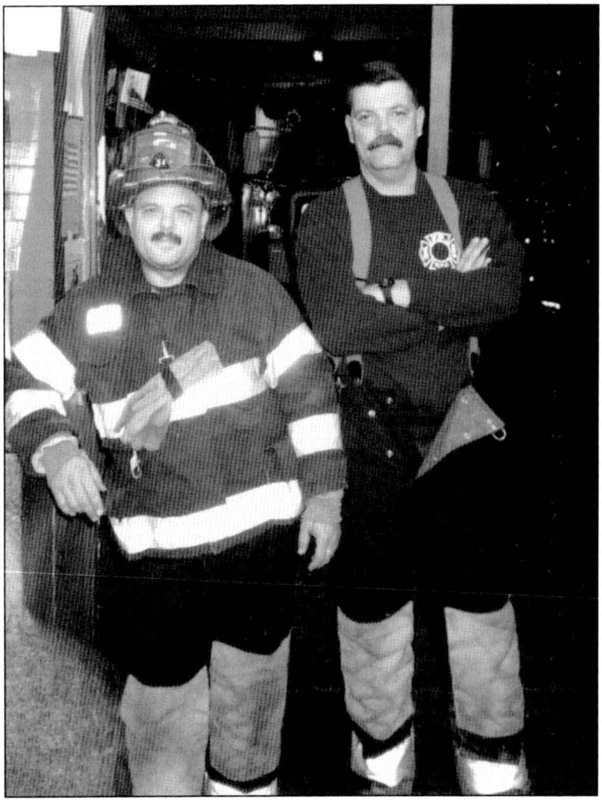

PATROLMEN JOE KRASKO AND MIKE STEMWEED. With a combined total of nearly 60 years of fire experience, these two friends are standing at the ready at Patrol 2. Patrolman Krasko is a longtime patrol historian and contributed greatly to this book.

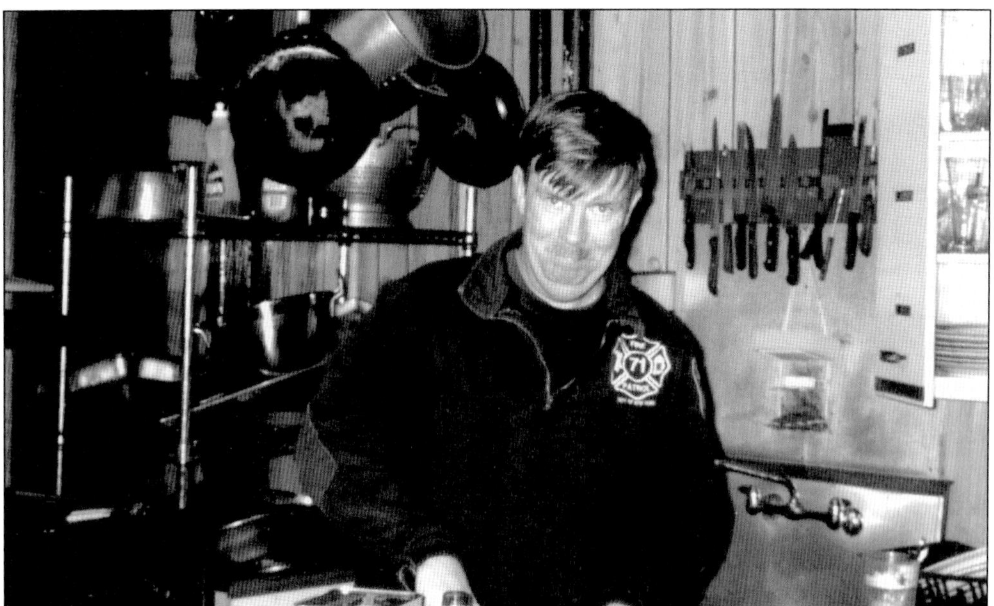

SENIOR MAN FRANK LENNON. Patrolman Lennon became a member of the patrol on November 5, 1971. He has attended thousands of fires in Manhattan during his 34 years (and counting) at Patrol 2. A chauffeur for many years, he is an avid patrol historian. This day, he took some time out to get a meal ready for the boys. (FL.)

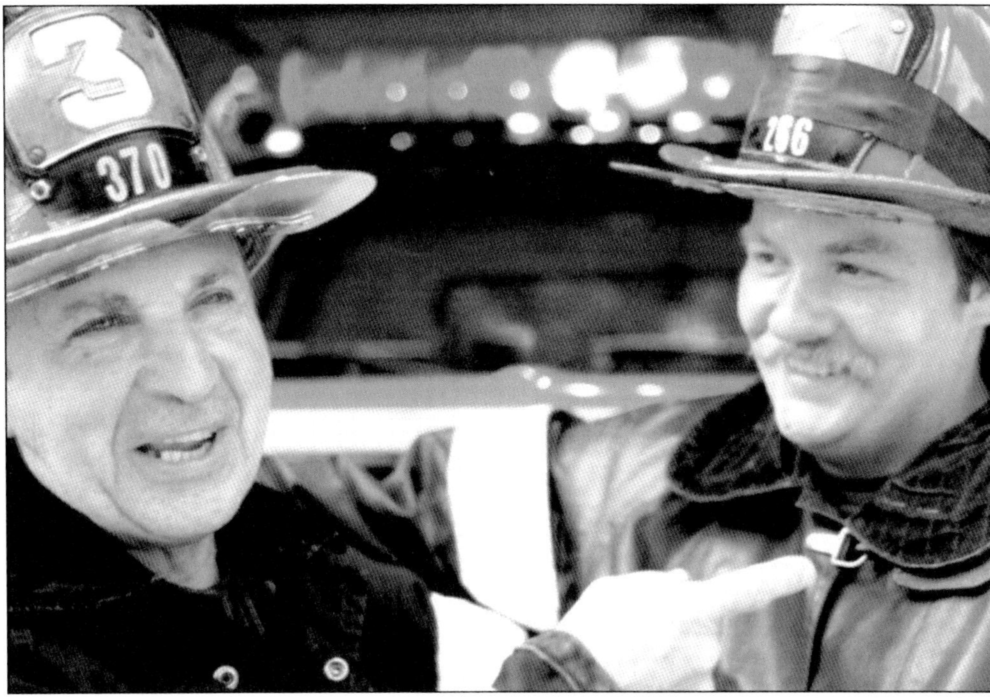

FRIENDS AT PATROL 3. Posing here for the camera is Al Savitsky and Frank Walters both of Brooklyn's Patrol 3. Savitsky had joined the patrol in 1952 and served for over 30 years. Walters was appointed on February 10, 1982, and assigned to Patrol 3. He was later promoted to sergeant of the shops, and is currently the patrol's safety officer. (DK.)

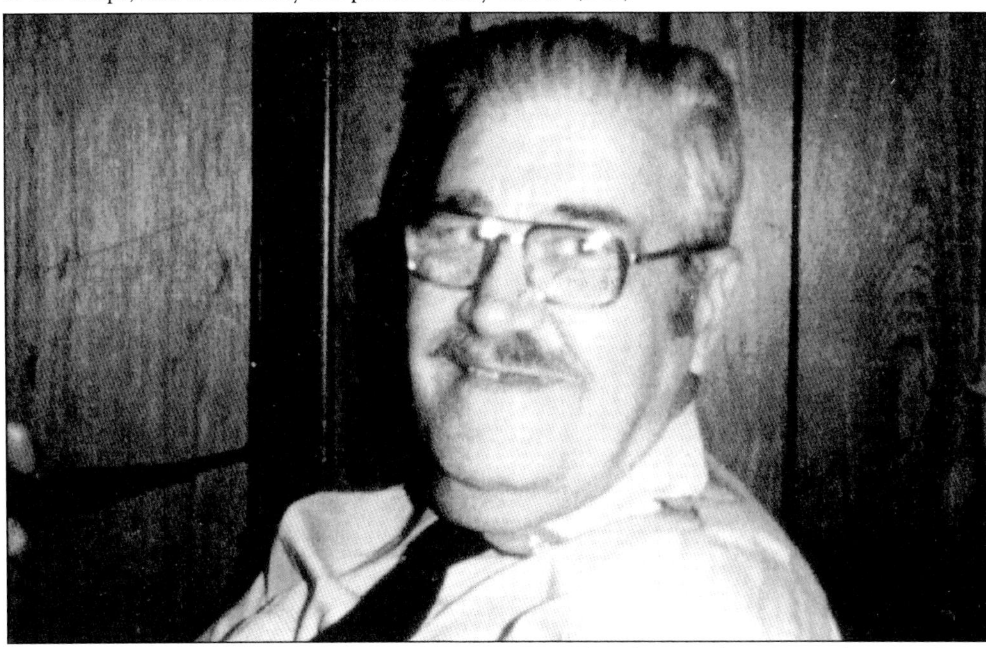

DEPUTY CHIEF LAVERY. Joseph P. Lavery was appointed to the patrol on September 4, 1951. He served in all ranks and retired as the deputy chief of the patrol in 1986. In a strange twist of fate, Joe Lavery passed away in 2001 and his funeral was held at 9:00 a.m. September 11, 2001.

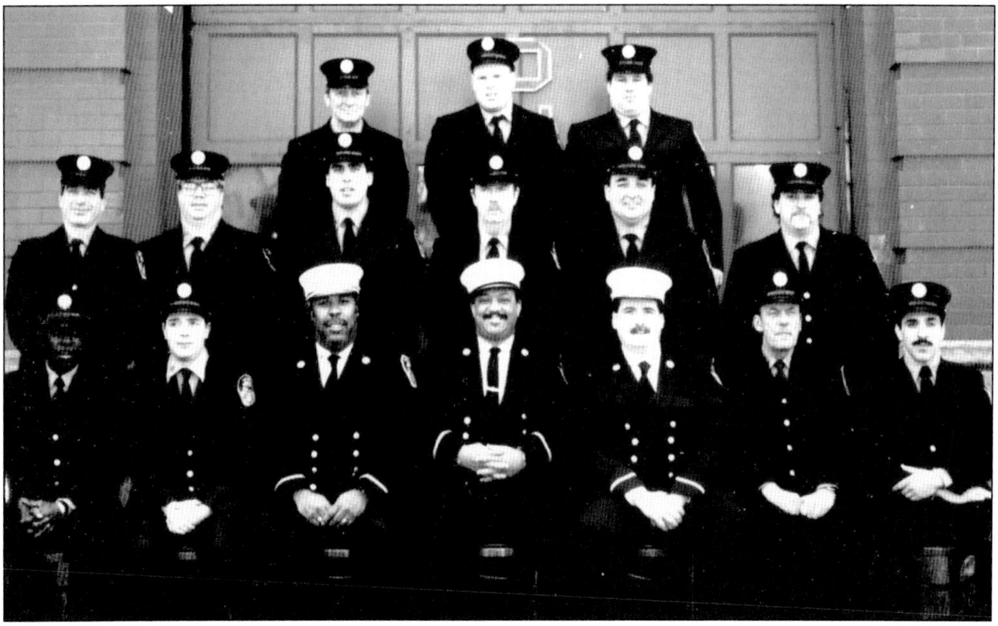

FIRE PATROL 2, 1985. Seen here, from left to right, are (first row) Emmanuel Reilly, Joe Crawley, Sgt. Milton Snipes, Lt. Rudy Adams, Sgt. John Sheehan, Bill Tobin, and Frank Boccia; (second row) Frank Tepedino, Jim McCauley, Frank Andrews, Frank Lennon, John LaCort, and Ed Leonard; (third row) Ed Fraunberg, Mike Trainor, and Jim Gorton. (FL.)

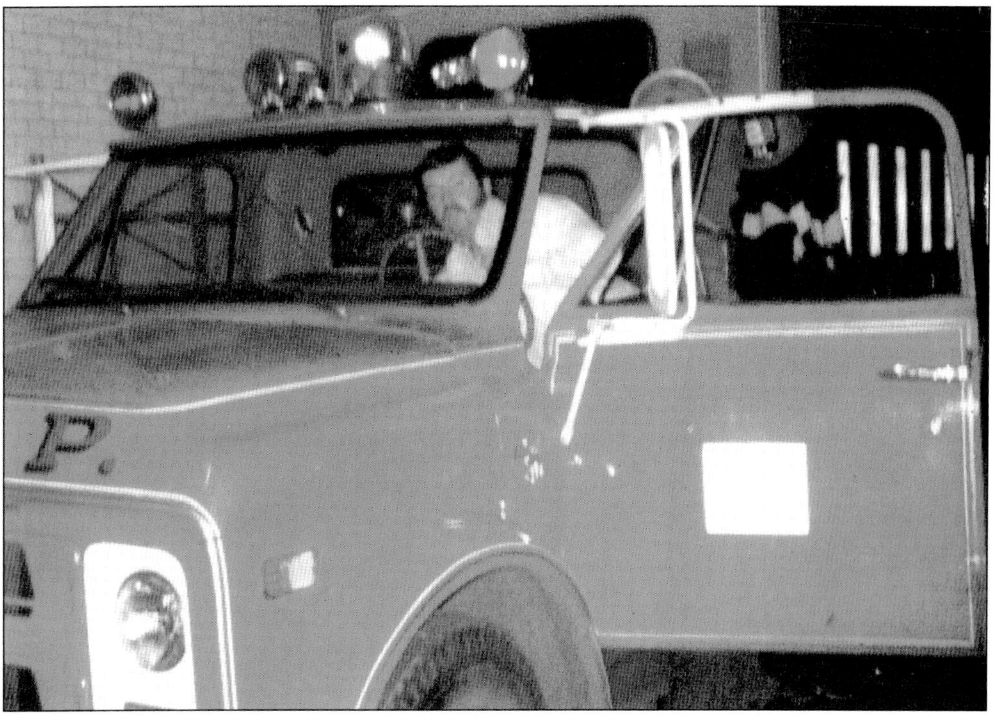

PATROL 2, 1971. John "Johnny" Brennan of Patrol 2 is shown here climbing behind the wheel of their 1968 Chevy. Brennan was a native of Brooklyn's Gerritsen Beach and served with Patrol 2 throughout the 1970s.

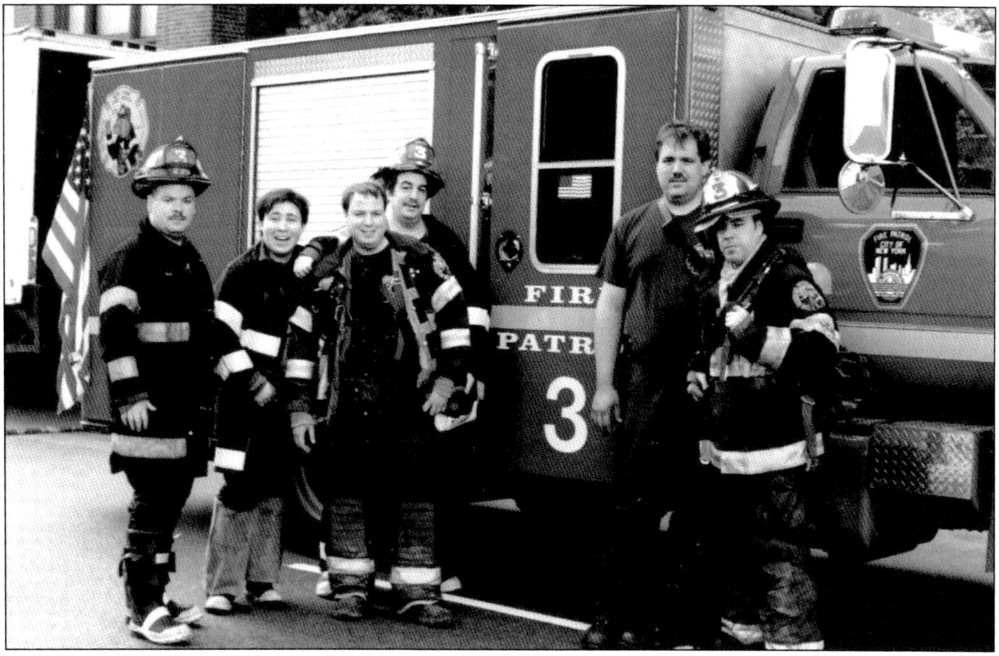

PATROL 3 DETAIL. This photograph by honorary battalion chief Jack Lerch was snapped at a transformer fire in downtown Brooklyn. From left to right are patrolmen Ronnie Stewart, Bill Foley, Larry O'Conner, Rich Lama, chauffeur Bob Kalb, and Sgt. Jim "Jimbo" Canham. (JL.)

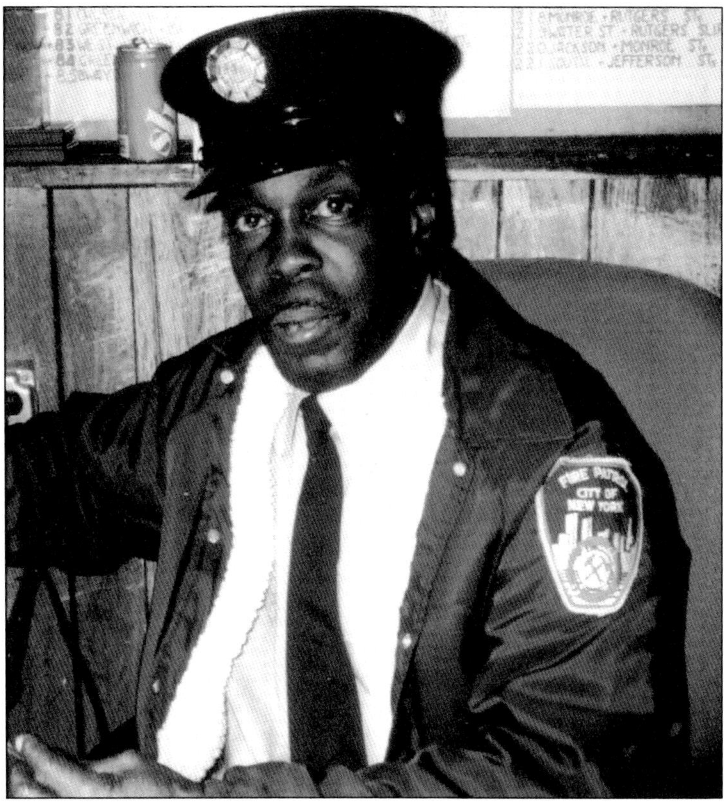

MANNY REILLY. Appointed on January 14, 1970, Manny "Moo Goo" Reilly was assigned to Patrol 2 for his entire career. An avid musician and steel drum builder, he was also known as the "only man who could eat a bowl of soup and drive a fire truck at the same time!" (FL.)

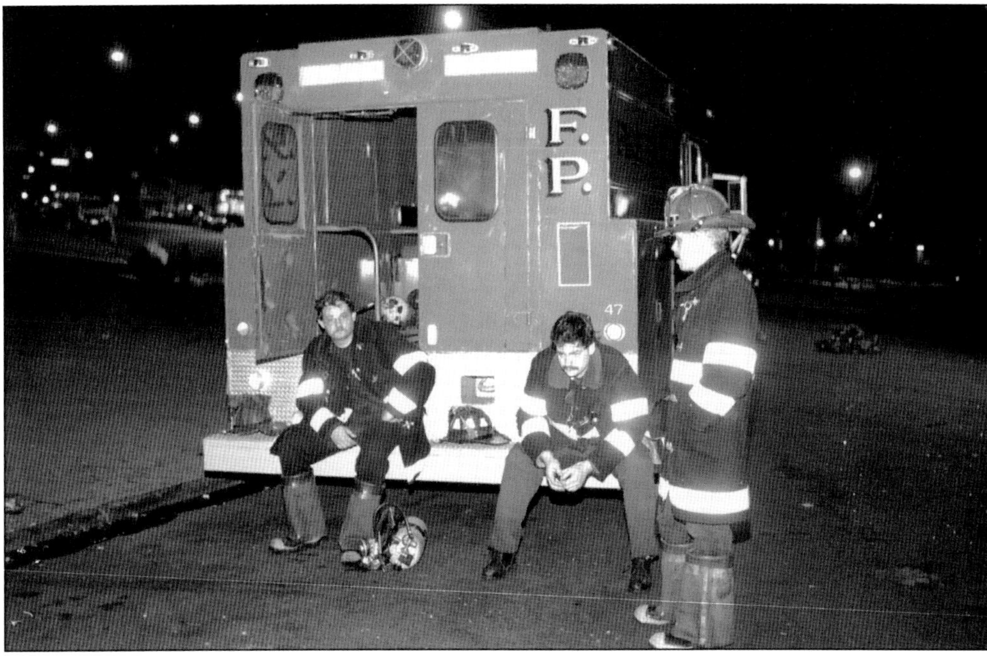

TAKING A BREATHER. The fire is darkening down, the merchandise has been covered, and the basement pumped. "Take a blow," shouts the officer and the men return to the rig to cool down. Here, patrolmen Ed Moran (left), Larry McGuiness, and Brian Mulvey (right) wait to take up at a Brooklyn fourth alarm. (BM.)

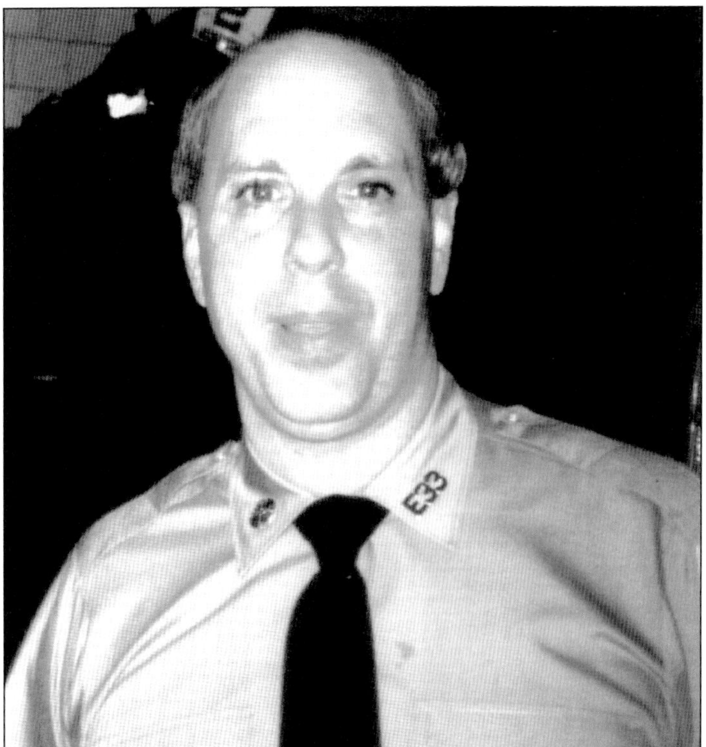

CARL OREGGIO. Carl Oreggio grew up across from Patrol 2, and from an early age watched the men of the patrol "turn out" from the old firehouse on West 3rd Street. In 1969, Oreggio was appointed a seasonal patrolman, a post which he held each summer with now NYFP chaplain Eugene Sawicki. Later, Oreggio joined the ranks of the fire alarm dispatchers, and eventually entered the ranks of the FDNY. Originally assigned to Engine 5 he later returned to the "Village" Engine 33, on Great Jones Street. Carl has been a familiar face at Patrol 2 for nearly half a century. (FL.)

FRANK R. TEPEDINO. Born in Brooklyn on November 23, 1947, Frank Tepedino made it big in 1967. It was during that year that he was drafted by the New York Yankees. During his career, the lefty from Brooklyn played mostly first base for the Yanks and also later for the Atlanta Braves. In 1975, Tepedino retired from baseball and soon after started a new career with no other than the New York Fire Patrol. Today he no longer wears the number 21 on his back, but the number 451 on his badge.

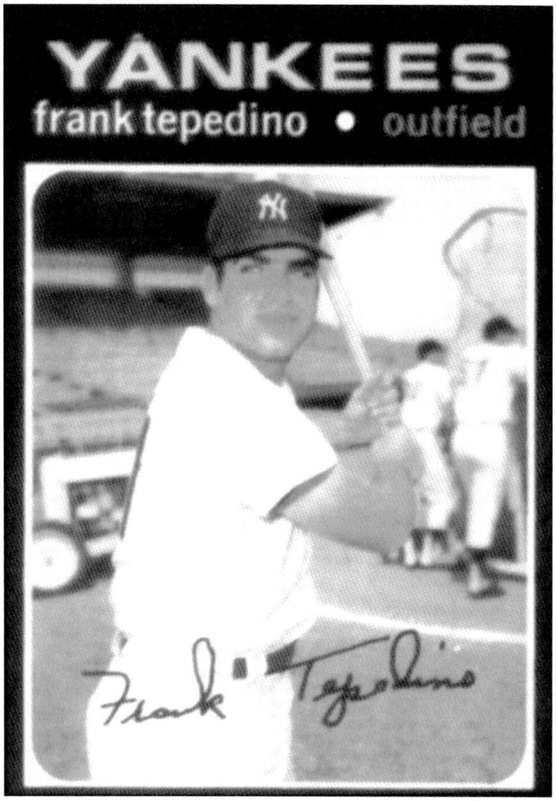

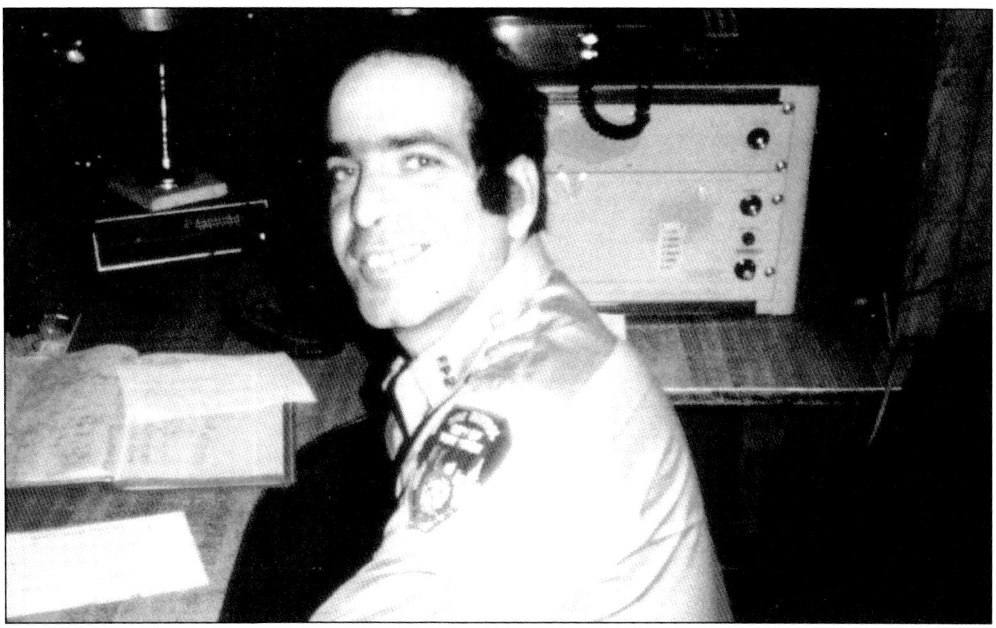

FRANK TEPEDINO. It's not everyday that you can get the chauffeur of Fire Patrol 2 to drive and autograph your baseball card of him, but in this case, it's okay. In Tepedino's years at the fire patrol, he has also served side by side with his two sons John and Frank Jr., both of whom spent time on the patrol. (FL.)

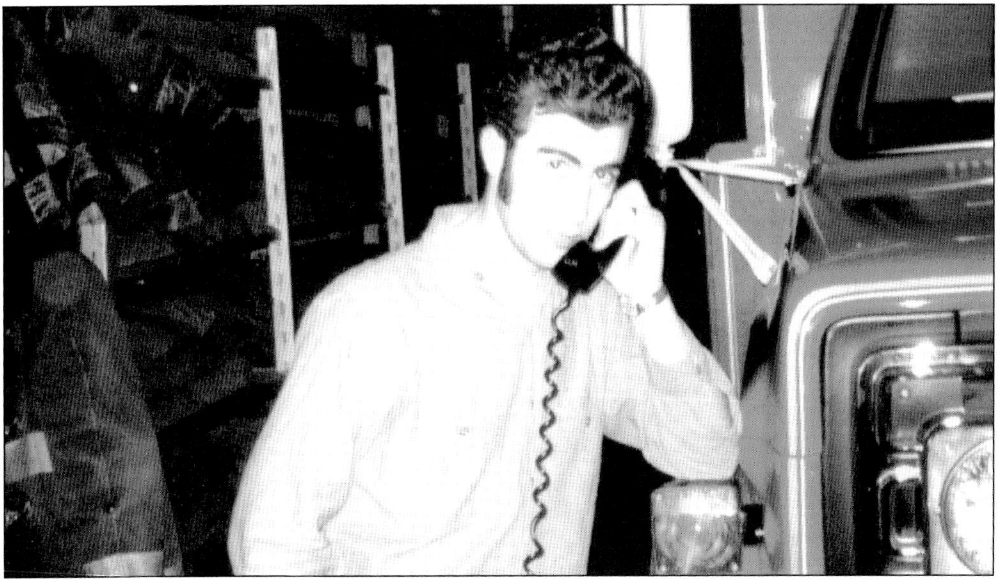

ARNIE ROMA, 1972. Arnold J. Roma was sworn in as a patrolman on August 8, 1972. Arnie, as most knew him, was assigned to Patrol 2 until he joined the ranks of the NYPD in the late 1970s. His son Keith would follow in his father's footsteps a quarter century later, and be assigned to the same house. (FL.)

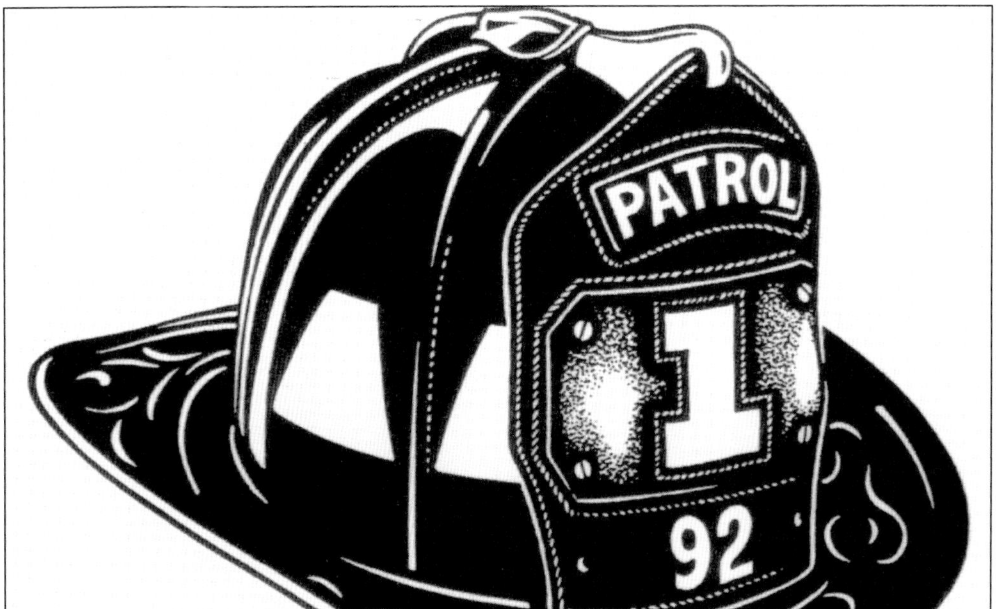

THE RED HELMET. The red fire helmet is the legendary symbol of the fire patrolman. When the red helmet first appeared is not know for sure. However, we know that in 1858, the following order appeared; "Resolved-That the men belonging to the Fire Patrol be required immediately at their own cost, to have their fire caps painted a bright red color, and have front pieces furnished thereto, with the words, "Insurance Patrol No. 1" for the lower district and "Insurance Patrol No. 2" for the upper district, together with the initials of the wearer of the cap legibly painted thereon to be white, with black letters as above." (SG.)

FIRE DEPARTMENT

CITY OF NEW YORK

UNIFORMED FORCE

DEPARTMENT
ORDER NO. 34 February 19, 1958

1.1 With regret, the deaths of

Sergeant *Michael G. McGee*, Fire Patrol
Patrolman *Louis J. Brusati*, Fire Patrol
Patrolman *James F. Devine*, Fire Patrol
Patrolman *Michael R. Tracy*, Fire Patrol

which resulted from injuries received while operating at Box 5-5-334, Manhattan, on February 14, 1958, is hereby announced to the Department.

The heartfelt sympathy of the entire Department is extended to the family, relatives and friends of the deceased in the midst of the great loss which they have sustained.

Funeral will take place on February 21st, 1958 and a 9:30 A. M. Requiem Mass will be celebrated at Holy Name R. C. Church, Prospect Park West and Prospect Avenue, Brooklyn, for Sergeant *McGee* and Patrolmen *Brusati* and *Devine*.

The funeral cortege of Patrolman *Tracy* will meet the Escort at point of assembly and participate in the funeral honors.

THE FOLLOWING FUNERAL HONORS ARE ORDERED:

The funeral escort shall be:

Fire Commissioner
Edw. F. Cavanagh, Jr.

Aide
Deputy Chief in Charge, *John J. Savage*

Deputy Fire Commissioners
George F. Mand *Harry P. Morr* *Albert S. Pacetta*

Secretary to Department
Denis Tilden Lynch

Chief Fire Marshal
Martin Scott

Acting Chief of Department
Arthur J. Massett

Aides
Deputy Chief in Charge *Richard A. Denahan*
Battalion Chief *Bernhard J. Muller*

STAFF

Assistant Chiefs of Department and
Deputy Chiefs in Charge on duty

Honorary Deputy Chiefs

Chaplains

Rev. Dr. Edward Lissman, *Rev. Merritt E. Yeager,*

DEPARTMENT ORDER NO. 34. When the signal 5-5-5-5 is given, all members are notified that there has been a line of duty death. Soon after, the fire department will issue orders stating the same. This order was issued to all city firehouses in 1958 upon the tragic Wooster Street fire in which four patrolman and two firefighters lost their lives.

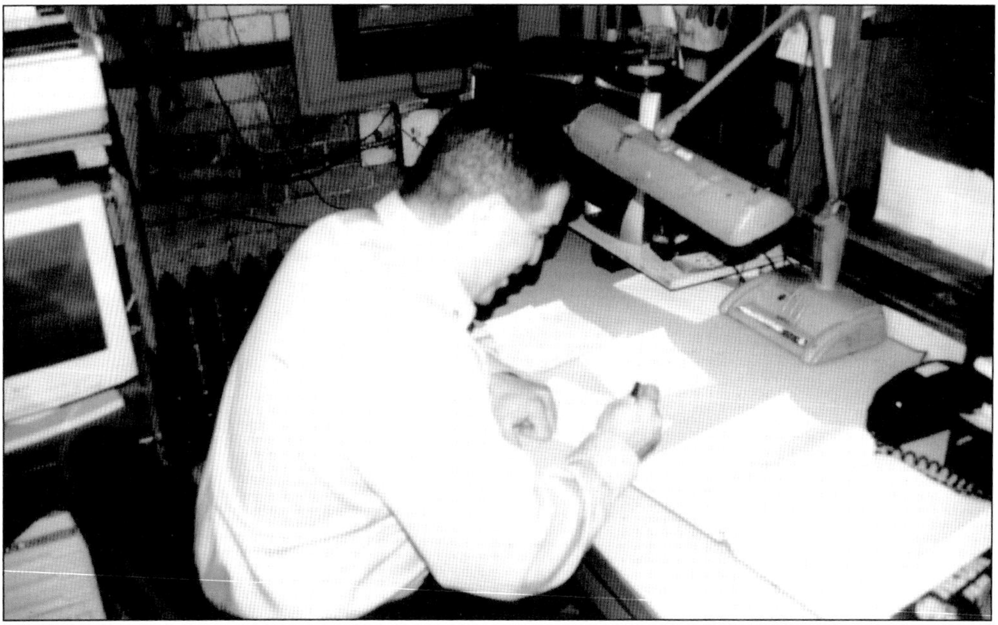

PATROLMAN ANGELINI ON WATCH. In a long standing tradition, the fire patrol housewatch man will remain behind when the detail responds to a fire. He will continue to monitor fires in the borough and keep order and security in the firehouse. Here, patrolman Mike Angelini completes the Form 1A, which must be filled out for every fire and emergency that occurs in the city.

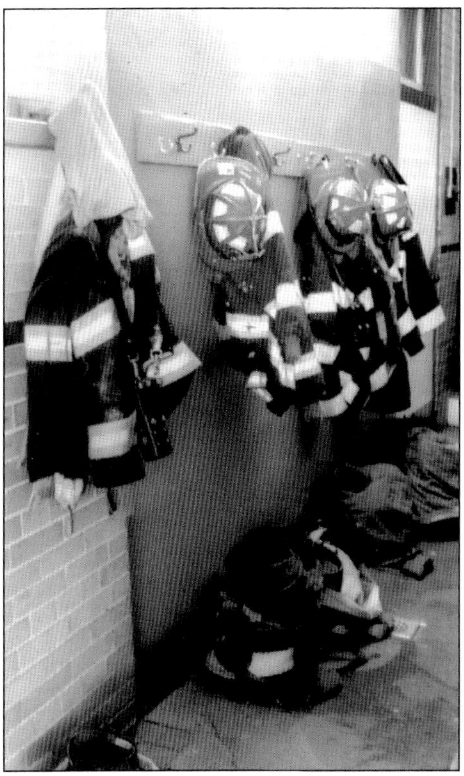

WAITING. Here the turnout gear of Patrol 3 hangs on the wall, waiting for the next alarm. For any fire service the thought of not returning to the firehouse after a run is unthinkable. Sadly for 32 members of the NYFP, they never returned to hang their red helmet and coat back on the wall.

Three

QUARTERS

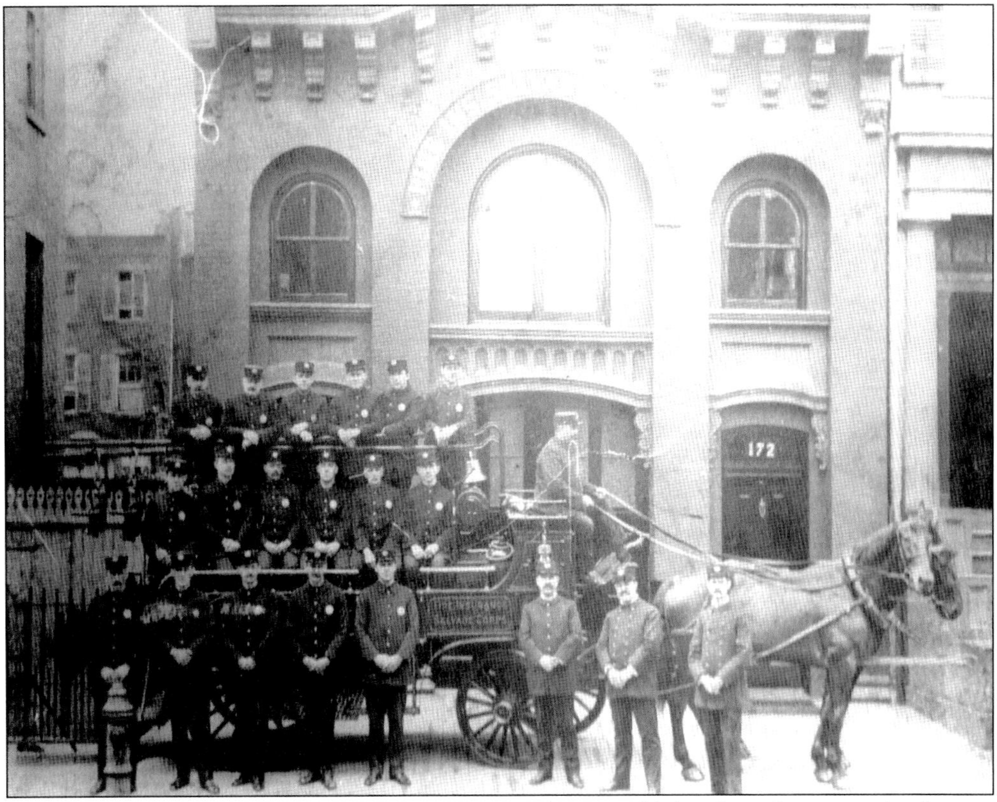

BFISC NO. 1. As early as 1880, an attempt was made to establish a fire salvage unit for the city of Brooklyn. However it was not until 1895 that it was officially begun. The new department was patterned exactly after the NYFP and began operations on December 16, 1895, from this renovated stable at 172 Pacific Street. The building stands today as a private residence. (LW.)

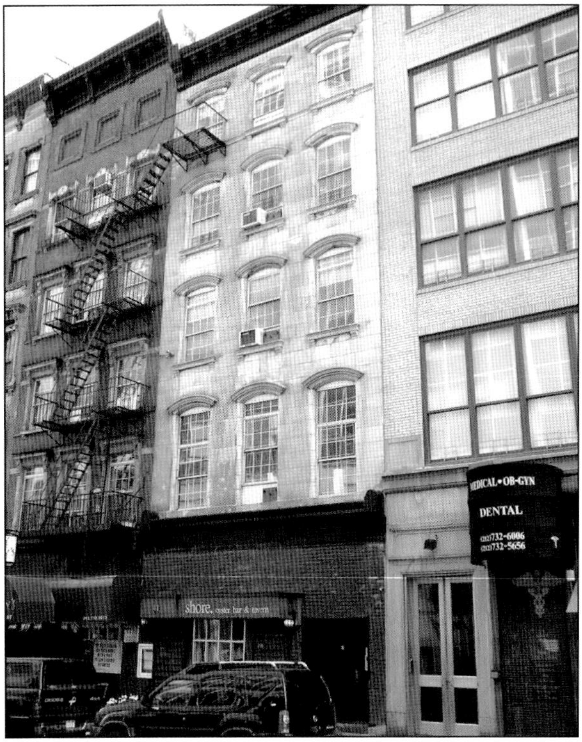

PATROL 1 QUARTERS. It has been written that as early as 1855 Patrol 1 was occupying quarters at 41 Murray Street. However, minutes of the patrol state no move to that location during that time. It is most likely that the patrol moved to this house in 1883. They remained here until moving to new quarters at 8 Franklin Street in 1923. (TER.)

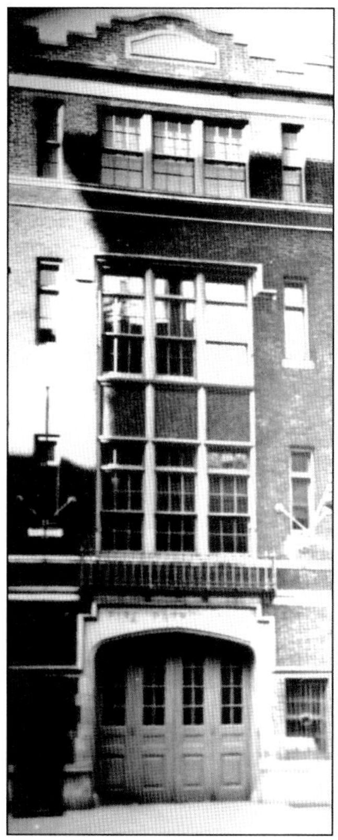

PATROL 1. After maintaining headquarters at 41 Murray Street, the NYBFU built this handsome station at 8 Franklin Street in 1923. Constructed in the art deco style, it was somewhat narrower then previous patrol houses. In use for only 14 years, the property was taken over for city redevelopment in 1937. Patrol 1 answered their last call at 5:26 p.m. on August 22, 1934. After this time, the 22 men assigned here were transferred to the other patrols throughout the city and the apparatus was assigned as the second section of Patrol 2. (PW.)

PATROL 2, 1866. Originally built for Engine Company 40 of the Volunteer Fire Department this building was located at 172 Elm Street. In 1866 Patrol 2 relocated here after it had been vacated by the new Metropolitan Fire Department. The building was renovated for patrol use and was used for a short nine months. In late 1867, No. 2 moved again to another former volunteer station about six blocks away. This engraving was featured in the 1880s book *Our Fireman*.

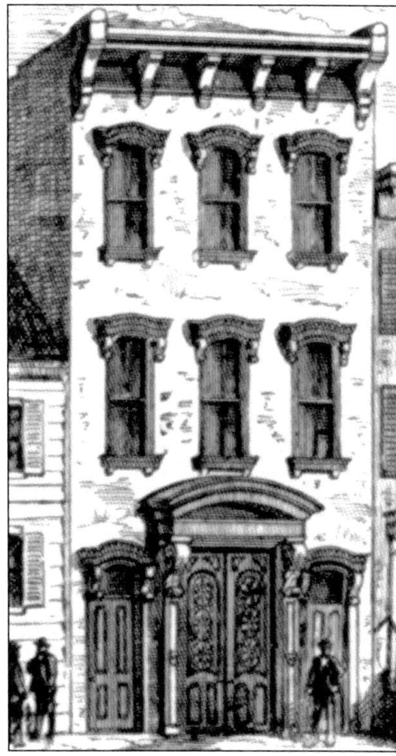

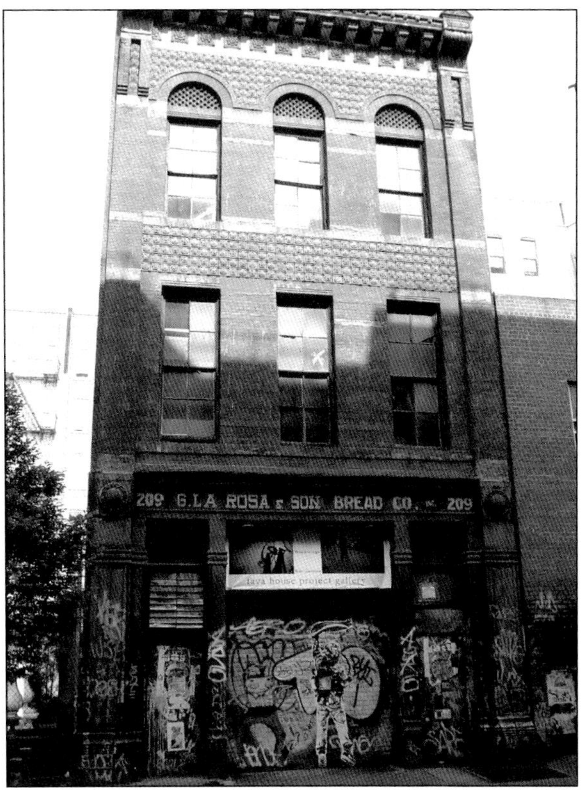

ELIZABETH STREET. This old firehouse at 209 Elizabeth Street was constructed for the Hibernian H and L Company 14 in 1863. It was vacated by that company upon the disbanding of the volunteer department in 1865. In 1866, Patrol 2 relocated to these quarters from Elm Street. They occupied this building for a few short years at which time it was turned back over to the city. It housed Ladder Company 9 until about 1950 at which time it was turned over to the Police Emergency Squad. (TER.)

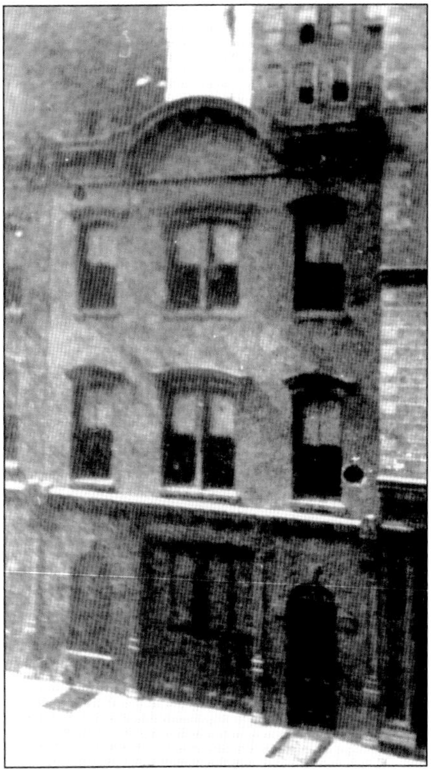

GREAT JONES STREET. Fire Patrol 2 moved around quite a bit as the New York Board of Fire Underwriters could not find a suitable piece of property to construct a firehouse. However in 1872, this large stable was found at 31 Great Jones Street near Broadway. It was then purchased by the NYBFU and converted into a suitable firehouse for Patrol 2. For the next 24 years they called it home. After the Fire Patrol vacated these quarters in 1907, the building once again became a commercial establishment. After many years of being used as a stable, garage, and finally vacant, the old firehouse started a new life as an upscale restaurant and apartment building. (Left FDNY, below TER.)

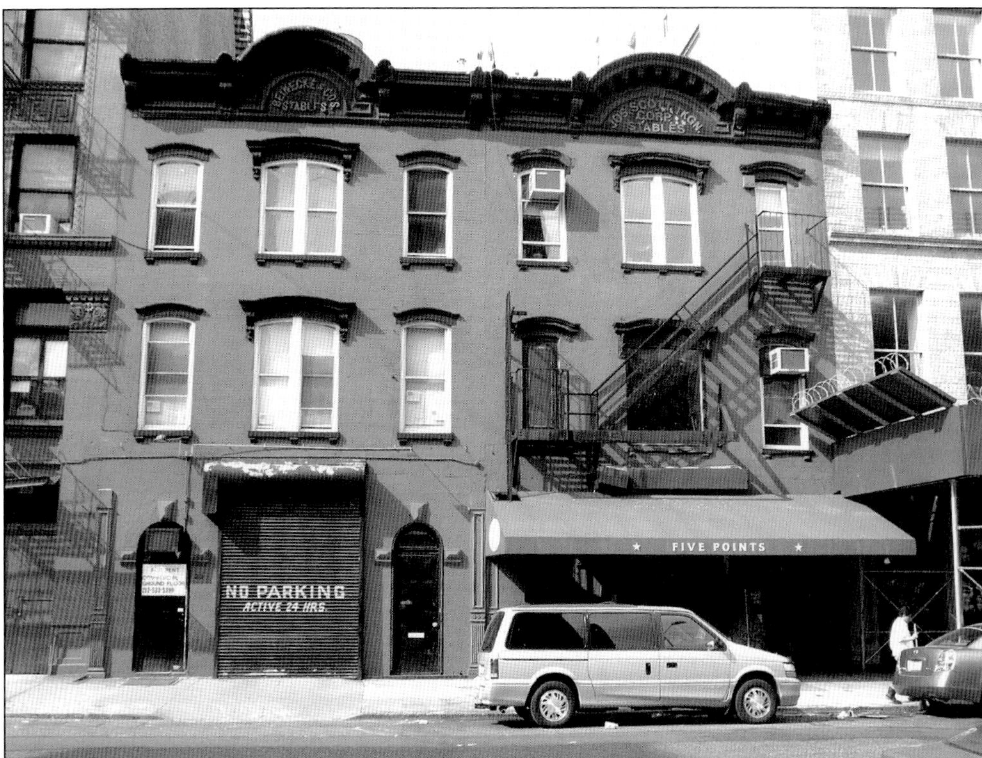

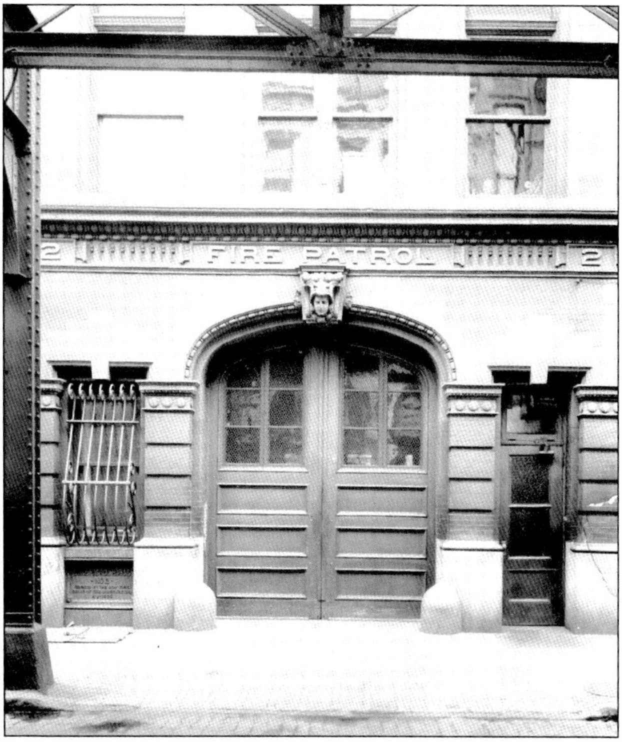

WEST 3RD STREET. After being in existence since 1855, Patrol 2 finally moved into a building to call their own. The NYBFU purchased this lot at 84 West 3rd Street, bounded by Thompson and Sullivan Streets in Manhattan's Greenwich Village. The new station was of the most modern design and contained a freight elevator to move the salvage covers to the drying racks on the fourth floor.

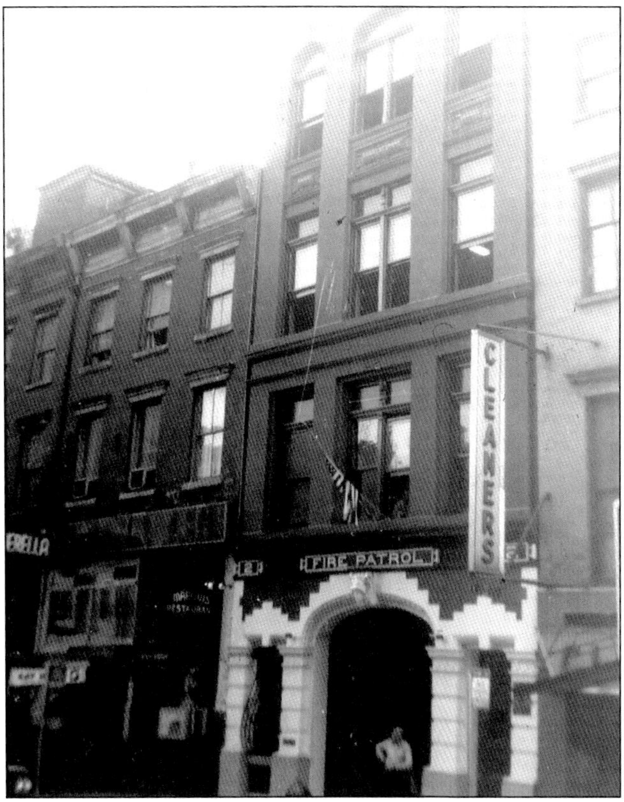

FIRE PATROL 2. It also contained a large apparatus floor and stables to the rear. Of note is the large chute located over the apparatus floor. At one time when the alarm sounded the driver would drop through a set double doors and land onto the seat of the patrol wagon. This cut down on response time and allowed the detail to hit the street within 30 seconds of the alarm. (FDNY.)

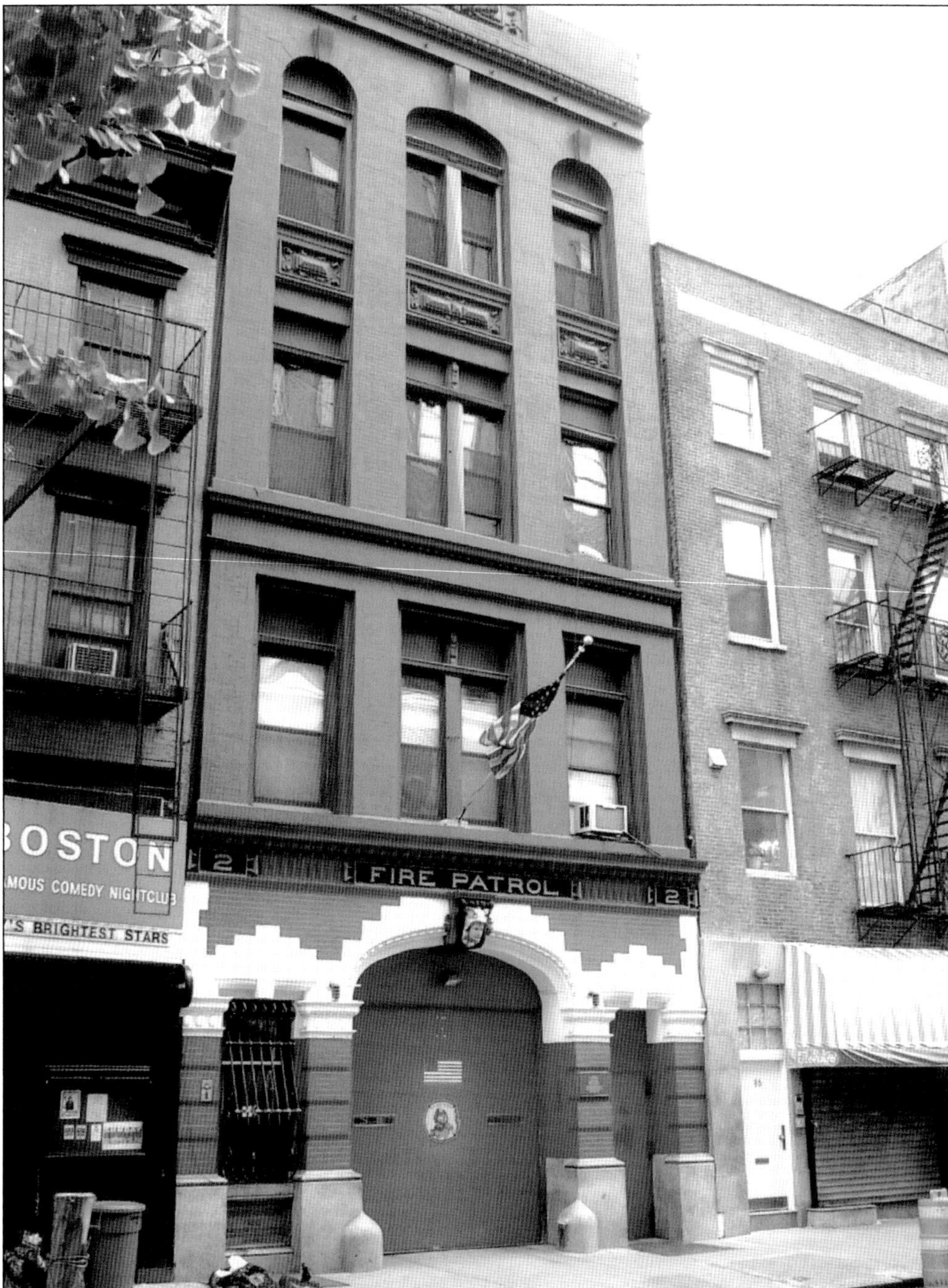

1908 Patrol House. This elaborate four story firehouse is of the classic "store front" firehouse design. Its single apparatus door is crowned with a bust of the Roman god Mercury, the winged god of speed and commerce. The building has been altered only slightly since 1907 and is a landmark in Manhattan's Greenwich Village. (TER.)

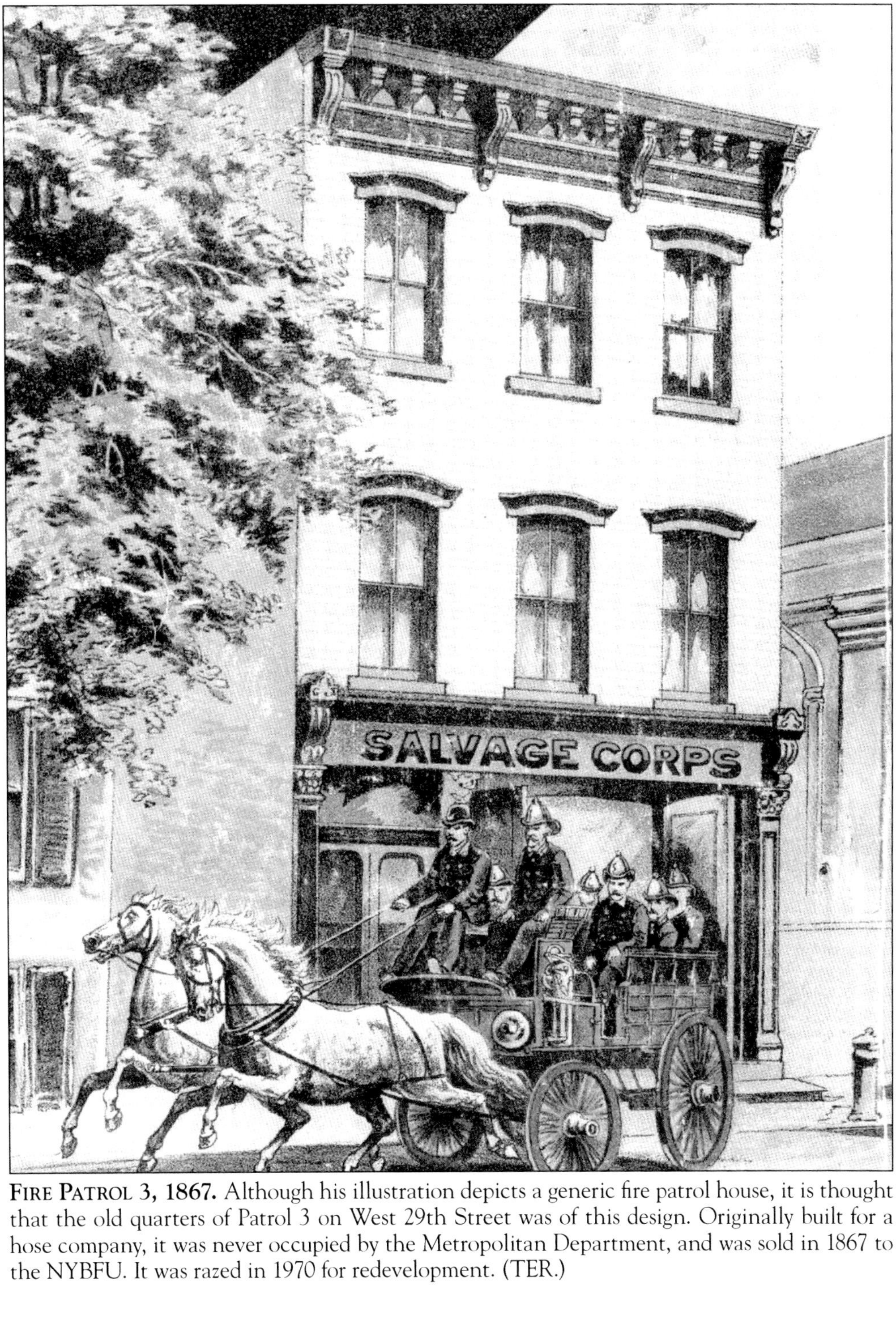

FIRE PATROL 3, 1867. Although his illustration depicts a generic fire patrol house, it is thought that the old quarters of Patrol 3 on West 29th Street was of this design. Originally built for a hose company, it was never occupied by the Metropolitan Department, and was sold in 1867 to the NYBFU. It was razed in 1970 for redevelopment. (TER.)

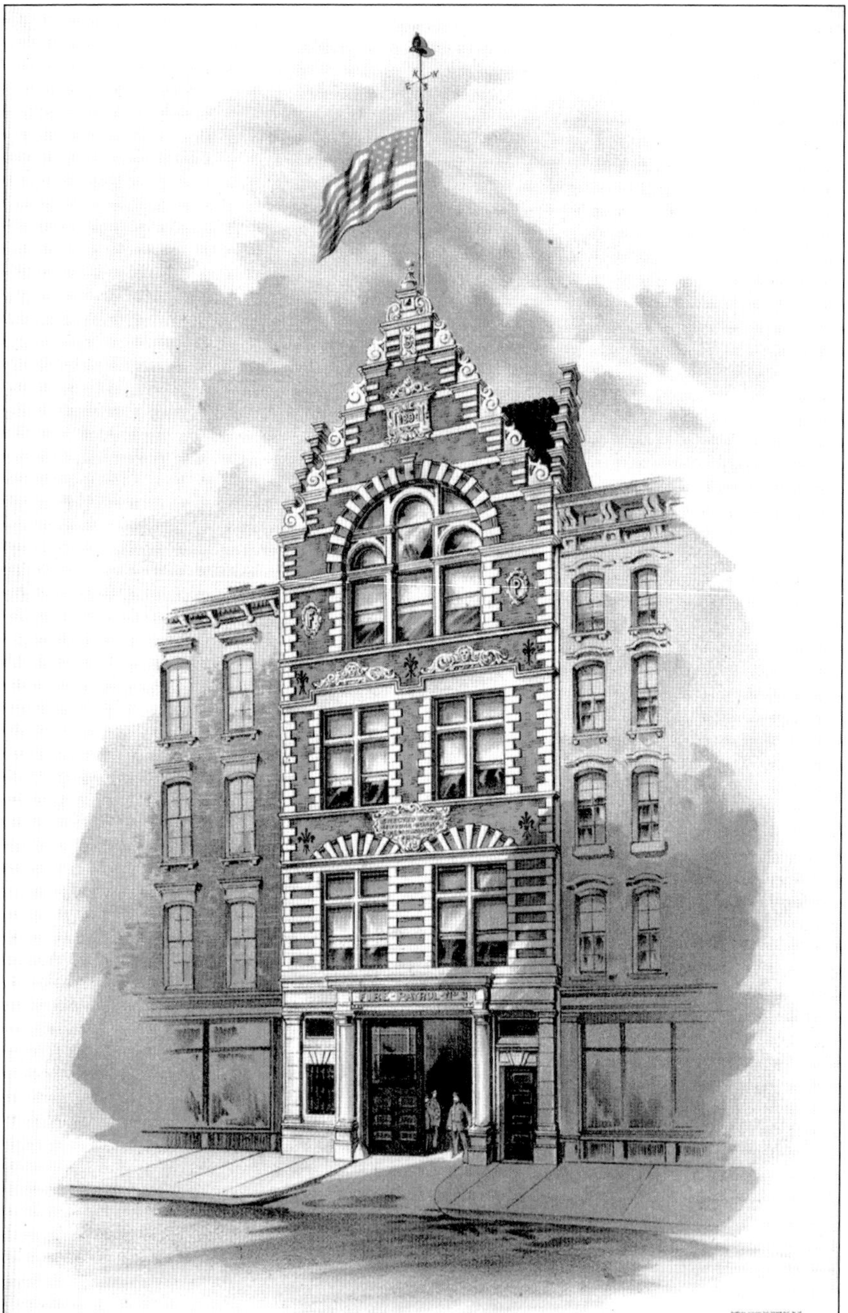

FIRE PATROL 3, 1895. One of the finest firehouses in the city was constructed in Flemish-style architecture of Indiana limestone. Located at 240 West 13th Street, this firehouse contained four floors, no less than 10 brass poles, a five horse stable, complete electric lights, and a fully plumbed washroom as well as the patrols repair facility and storerooms. Occupied by Patrol 1 under the command of Capt. Peter Cornwell, his detail moved here during the week of September 10, 1895, from their old quarters located two blocks away at 104 West 13th Street. (LW.)

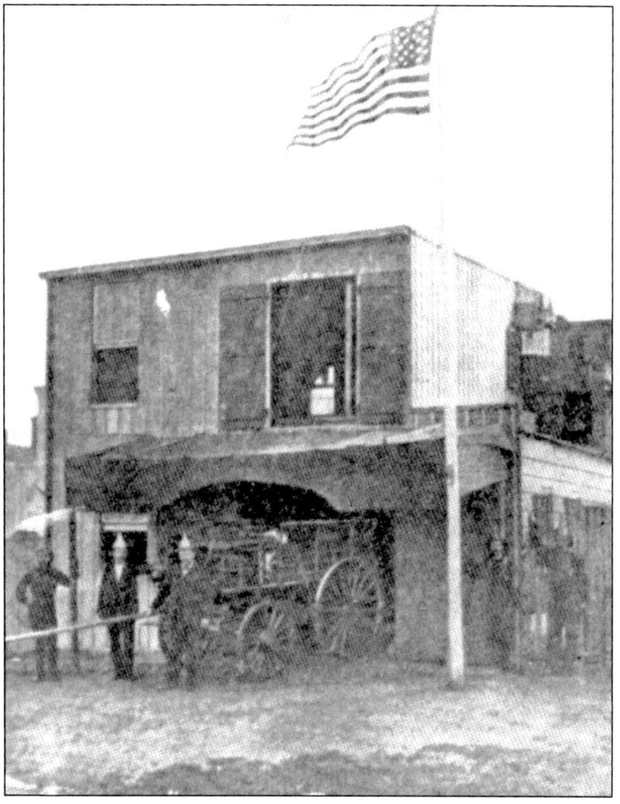

PATROL 4. Located at East 94th Street and Lexington Avenue, the patrol moved into this crude metal building upon its formation in 1876. At this time, they were known as the Eastside Patrol and responded to all alarm boxes in Manhattan to about 120th Street. The patrol wagon in this photograph is of unusual design and is apparently an older wagon which had been a spare, until being pressed into service with the new Eastside Patrol. They would remain here until 1883. (NYFP.)

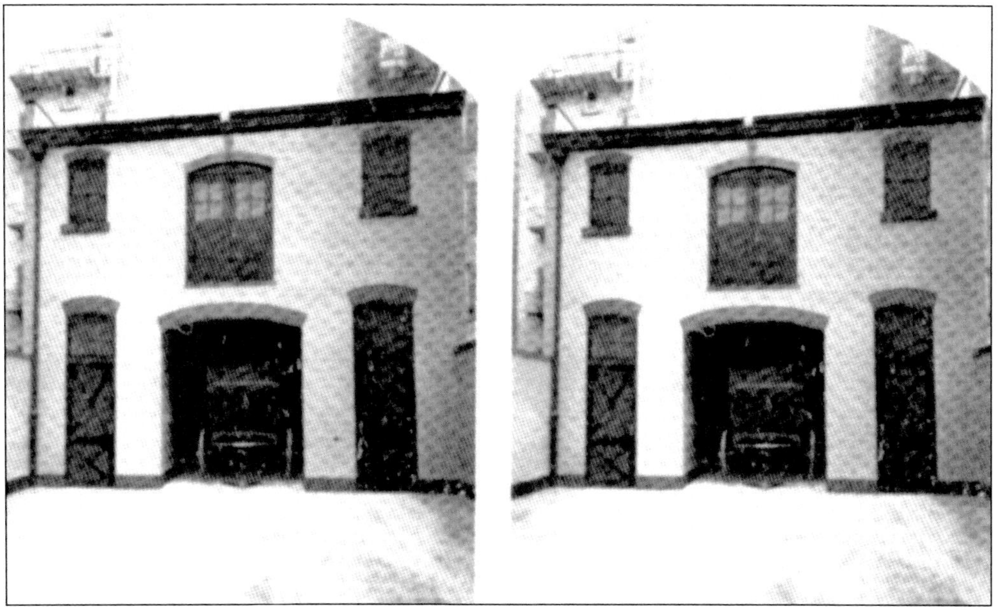

STABLES AT PATROL 4. Aside from being the home for this company's horses, one of the spare wagons was kept here. This elaborate two story stable also had room for the forge for shoeing horses and a hayloft above. In later years it would be used to store extra equipment and supplies for the Fire Patrol School of Instruction. (LW.)

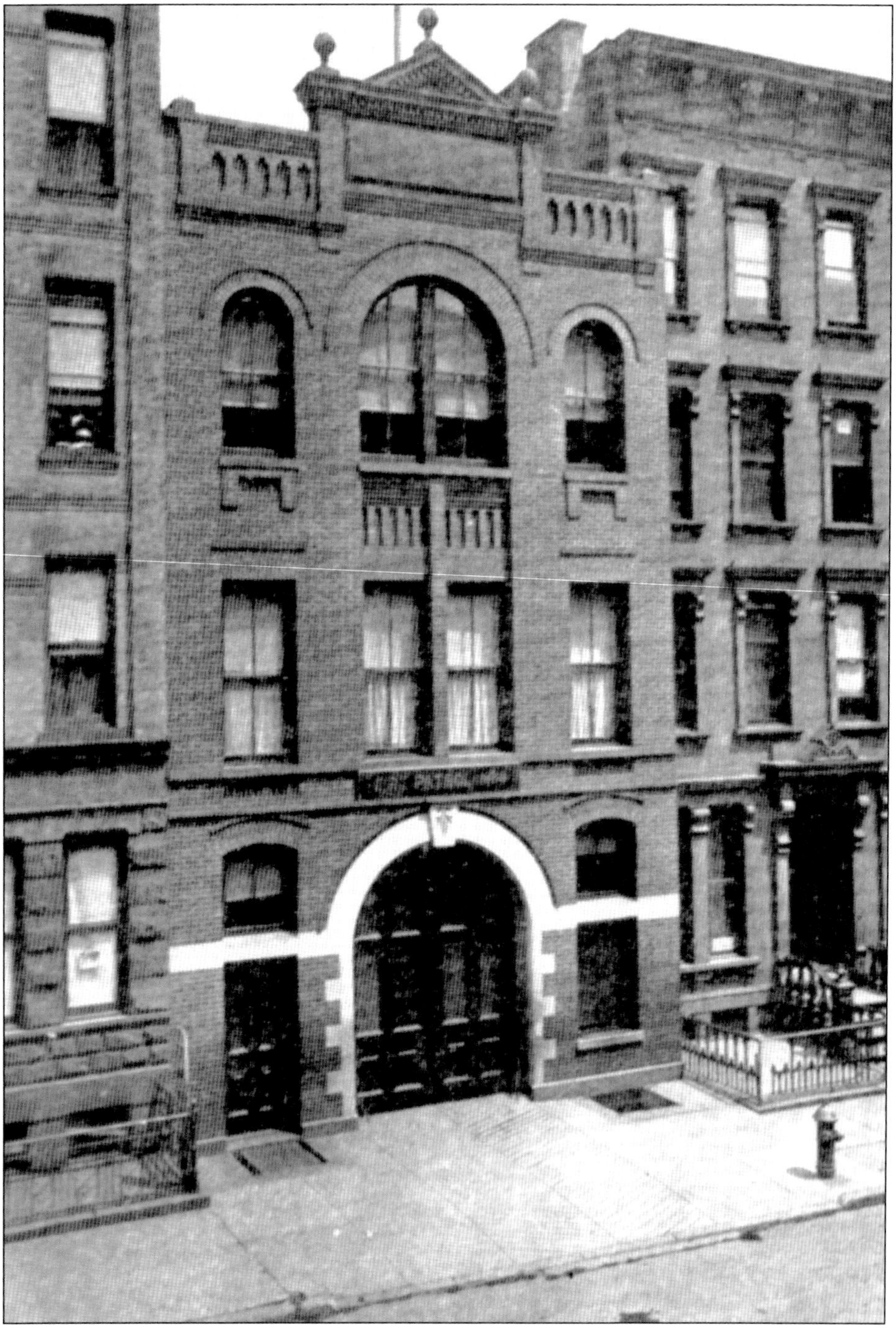

New Patrol 4 House. Early in 1882, the NYBFU purchased a lot on East 90th Street between Madison and Park Avenues to erect permanent quarters for Patrol 4. In May of 1883, the company left the old tin shack on East 94th Street and moved into this handsome firehouse. They would call this building home until they were disbanded on July 1, 1941, due to fiscal constraints. At this time, Patrols 1 and 5 expanded their coverage into this area. It stands today, unchanged, as a private residence. (NYFP.)

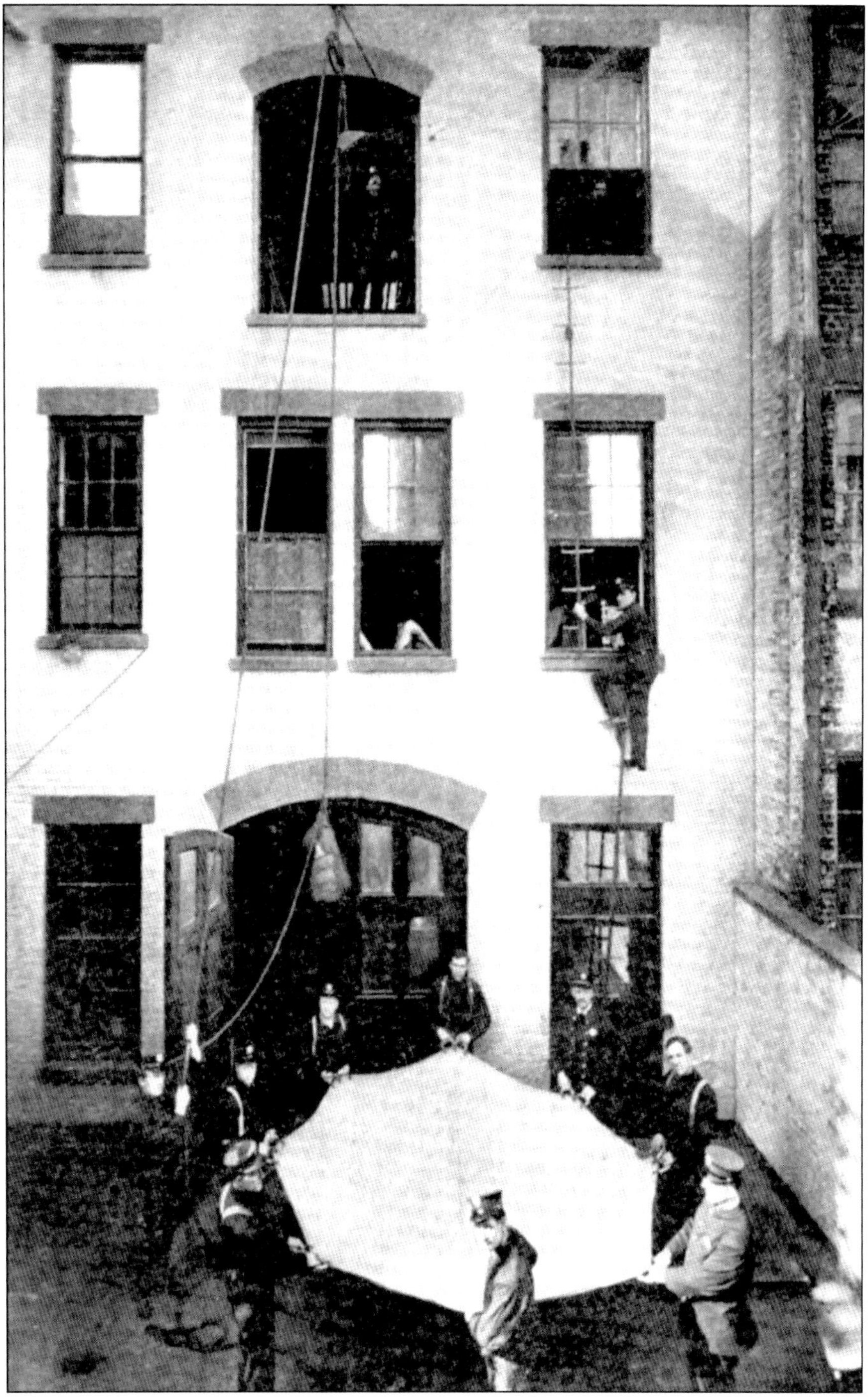

DRILLING, 1930. This view is looking toward the rear of the quarters of Patrol 4 from the top of their stables. Here the company is holding classes on ropes, scaling ladders, and the life net. It is unknown when the patrol carried life nets, but the scaling ladders were in use until about 1940. The Fire Patrol School of Instruction was conducted here for many years and all probationary patrolman were instructed here in all aspects of salvage operations. (JK.)

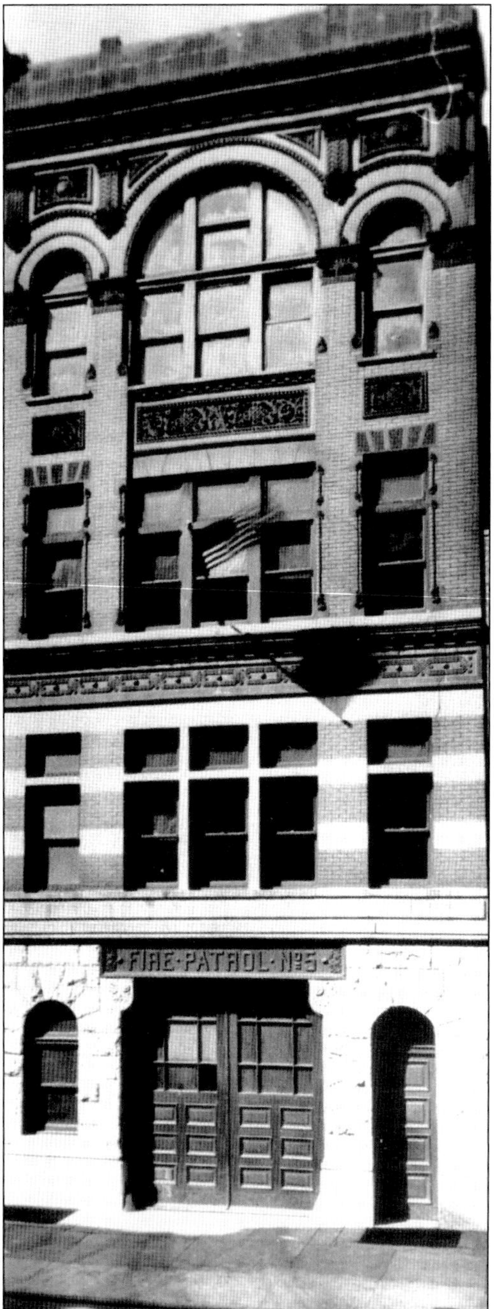

A NEW HOUSE IN HARLEM. Up until 1892, coverage for areas above Central Park were handled by Patrol 4. Due to the tremendous growth of the area above 110th Street, the NYBFU decided to open a fifth patrol company in the predominantly German, Irish, and Italian neighborhoods of Harlem. The company began service on February 1, 1892 in this beautiful four-story brick and limestone firehouse located at 309 West 121st Street. One of the busiest companies in the city, Patrol 5 covered all of Northern Manhattan to the Bronx Borough line. On November 1, 1954, the company was closed and the building sold at auction. (PW.)

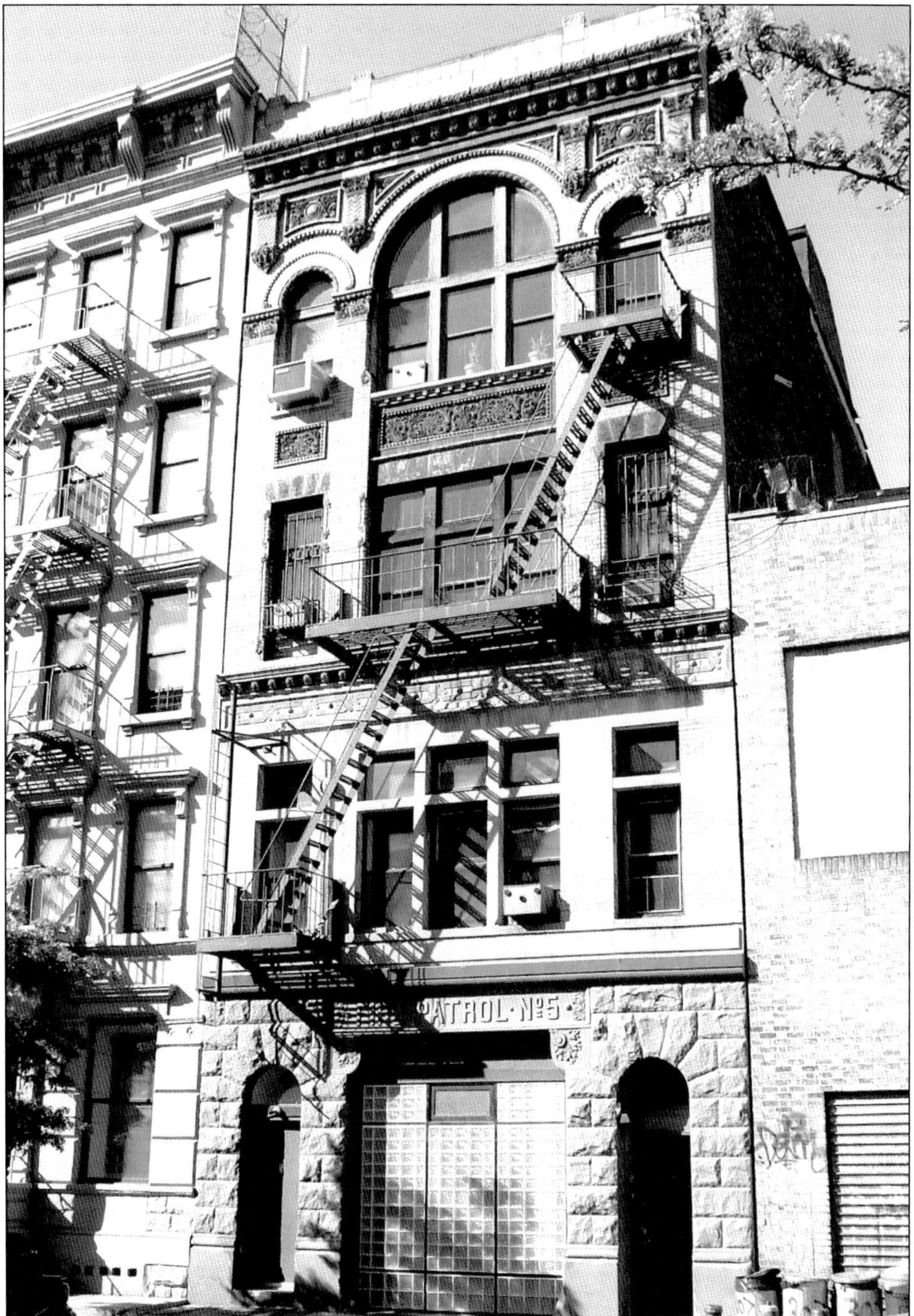

PATROL 5. "After a careful study of the operations of the Fire Patrol, the member companies supporting the Fire Patrol voted unanimously to discontinue Patrol 5. . . . " As it was considered, the judgment of the members was that the concentration of values in the district no longer warranted the expenses incurred in providing protection by the fire patrol. During the year of its closing in 1954, the patrol had answered 16,555 alarms and spread over 21,000 covers at fires. It has been an apartment building since its closing in 1954. (TER.)

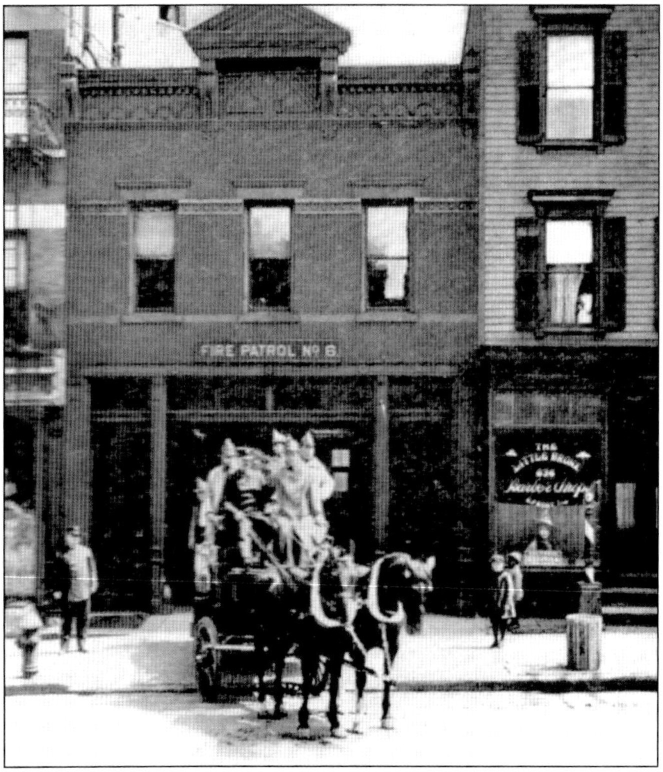

UP IN THE BRONX. Fire Patrol 6 began responding from 838 Courtland Avenue on December 2, 1901, in the Melrose Section of the Bronx. During this time the population of the Bronx was fast approaching 100,000 and Patrol 6 would soon have their work cut out for them. They were housed in this building for just 12 years. The building stood until 2004 when it was torn down for urban renewal. (JK.)

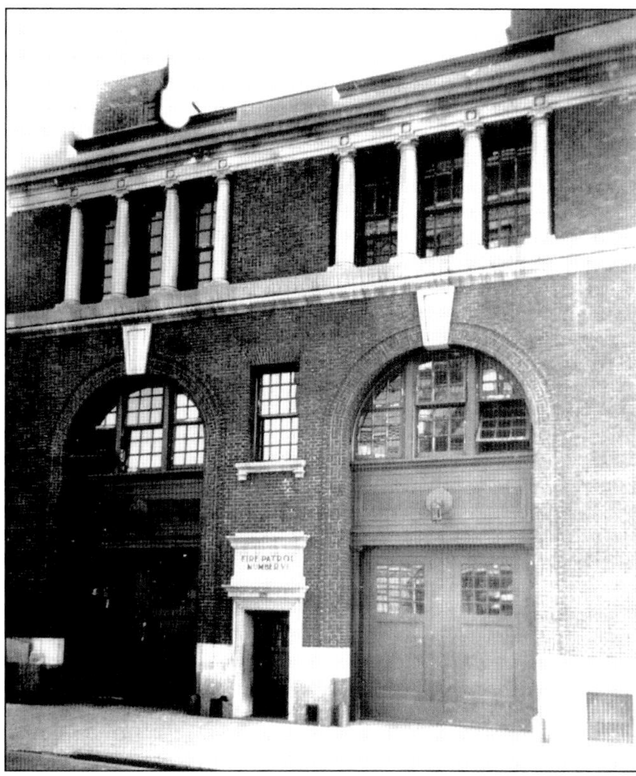

NEW PATROL 6. The most substantial of all fire patrol houses was completed under the direction of Superintendent James O. Schwank in 1913. It remains the only house erected by the patrol after motorization and is the only building not to have horse stalls. In 1934, the training school was relocated here under command of Capt. William Cashman. Just 36 years after it was completed, the property was purchased by the city and was razed for public housing. (JK.)

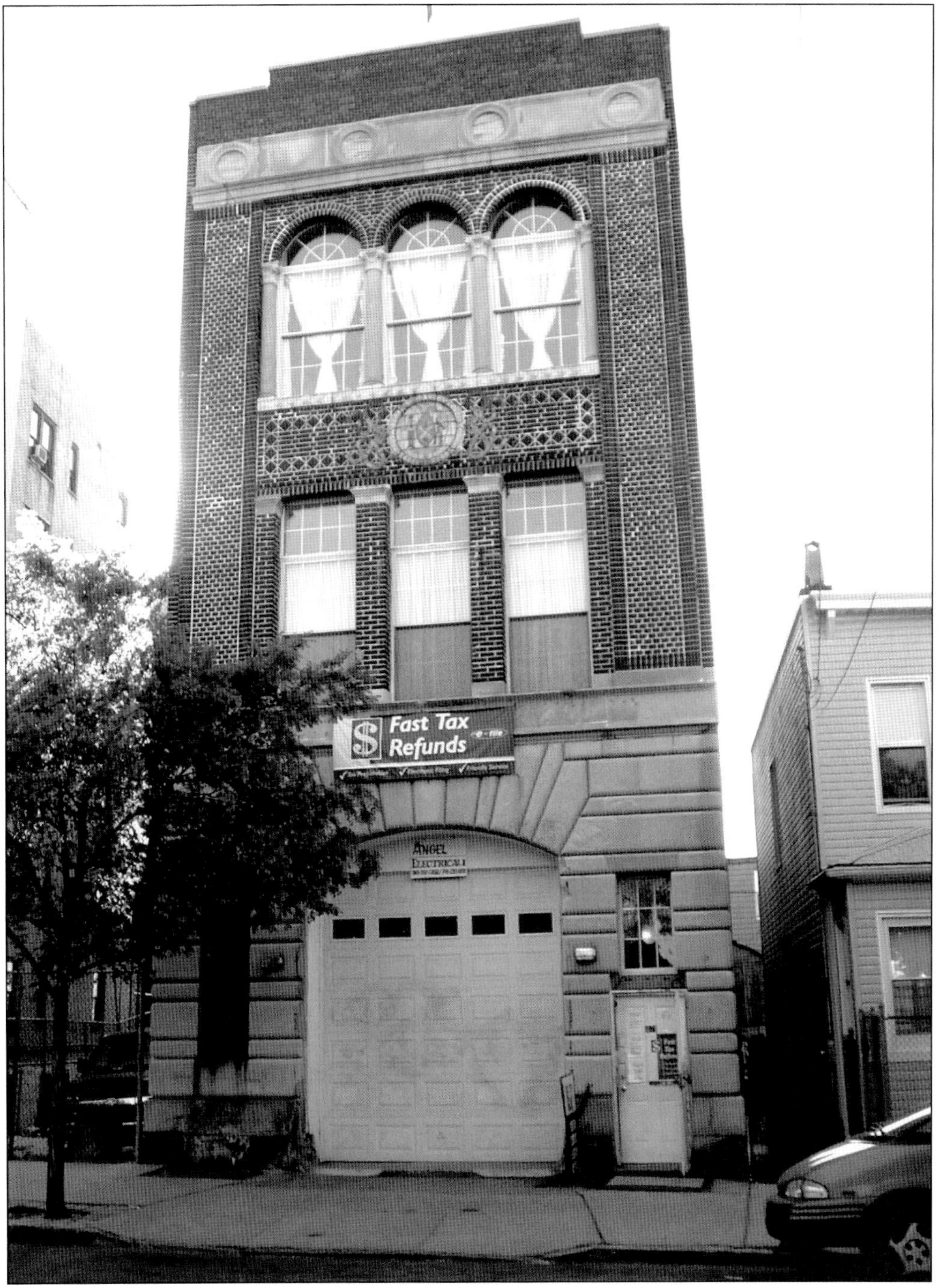

ANOTHER NEW HOME. Forced to find new quarters for Patrol 6, the NYBFU was able to purchase this former FDNY firehouse at 1079 Nelson Avenue in the High Bridge section of the Bronx. It had served as the quarters of Ladder 49 from 1913 until 1947 when they moved to 1080 Ogden Avenue, directly behind this building. On July 1, 1955, Patrol 6 was renumbered Patrol 5, and were disbanded just a few months later on December 5, 1955. After sitting vacant for nearly 30 years, the building has been renovated and returned to its former splendor. (TER.)

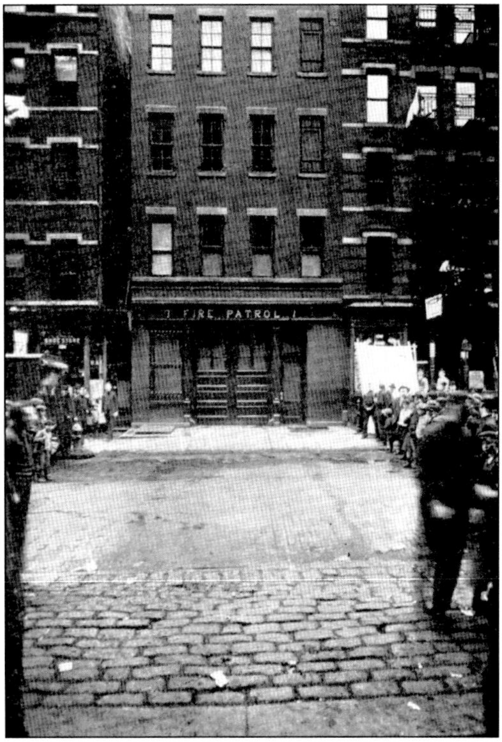

FIRE PATROL 7. Located in the heart of Manhattan's Lower Eastside, Patrol 7 was organized on September 3, 1906, at 133 Norfolk Street. The firehouse had been a former tenement and was converted by the underwriters into a modern fire patrol house. Four stories high, it was surrounded by one of the most diverse and crowded urban areas in America. On the patrol's block alone, nearly 1,000 people lived in crowded tenements. (NYFP.)

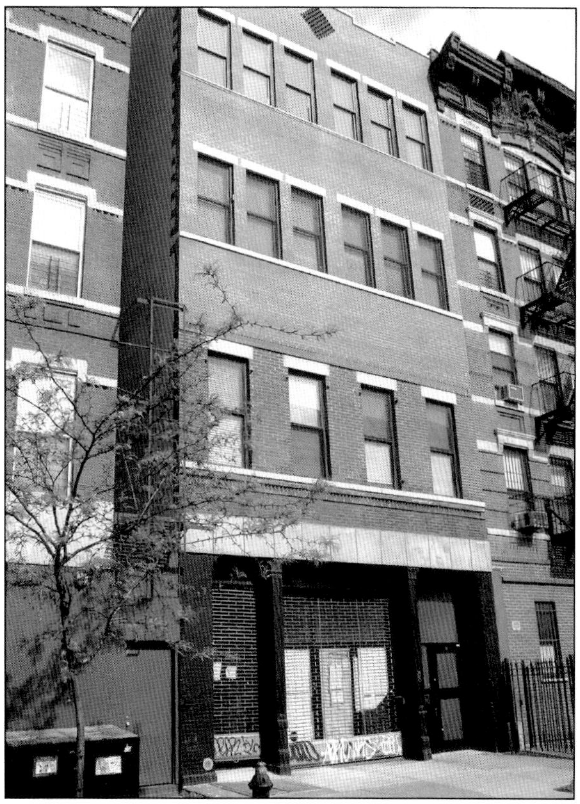

PATROL 7. The death knell for Patrol 7 was sounded upon the crash of the stock market in 1929. Patrol 7 was closed on September 30, 1933, and its equipment and men were transferred to other houses. In the summer of 2002, Timothy Regan (author), under command of Captain McGee and patrolmen Galloway, Fink, and Isakson responded to an "all hands fire" in the old firehouse, and threw 22 covers in the historic building. (TER.)

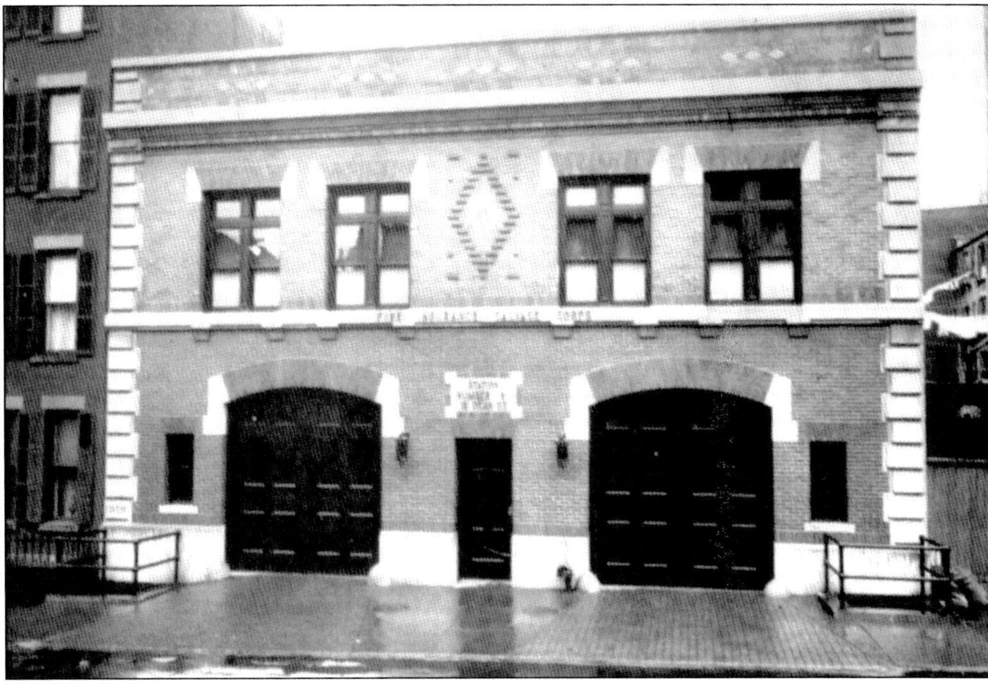

PATROL 8. Located at 10–12 Dean Street in Brooklyn's Boreum Hill area, this building was the first two-bay firehouse constructed for the patrol, and one of the first in the entire city. Originally built as the home of the Brooklyn Fire Insurance Salvage Corps 1, it was renumbered as Station No. 8 of the NYFP upon consolidation in 1911. From 1911 until 1955, it was known as Station 8, at which time it became Patrol 3. (MK.)

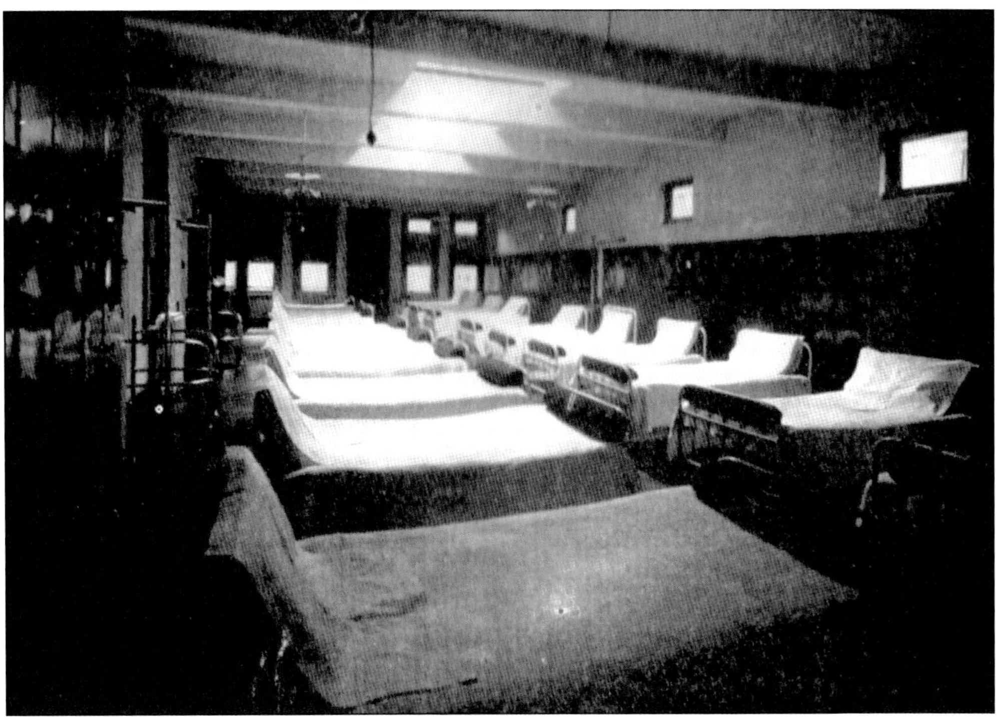

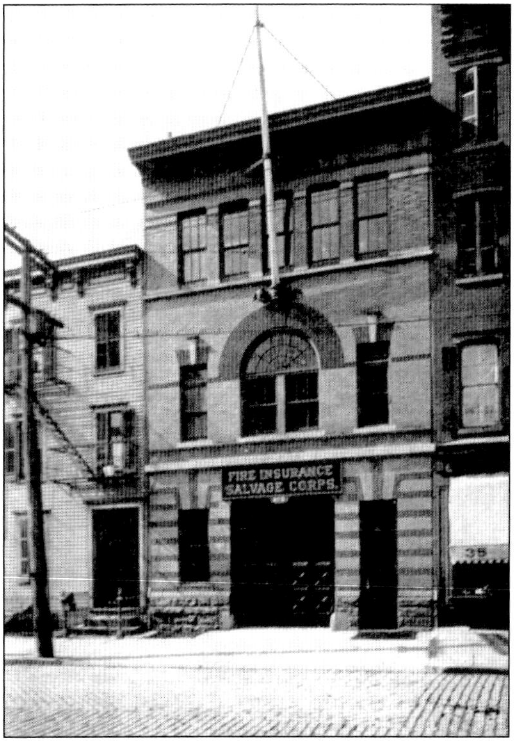

PATROL 9. Designed by R. L. Daus, this fine building was completed in 1896 for the BFISC at 33 Stagg Street. This three-story building contained over 4,700 square feet of working and living space. It contained stalls for four horses and two patrol wagons, as well as sleeping quarters for 15 men. From this station the men protected all of the large industrial and commercial area of Greenpoint in Brooklyn.

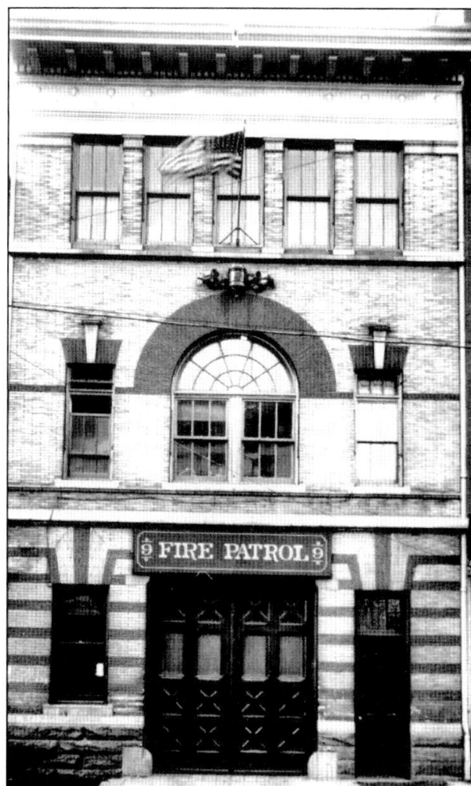

PATROL 9. In 1911, this company was redesignated Fire Patrol 9 of the NYFP, and in 1955, it was redesignated once again as Patrol 4. In May 1959, Patrol 4 was closed during its last year of operation. Patrol 4 made up part of the $5,381,677.11 savings from fire damage on insured property. During this time the patrol cost just over $600,000 to operate. (LW.)

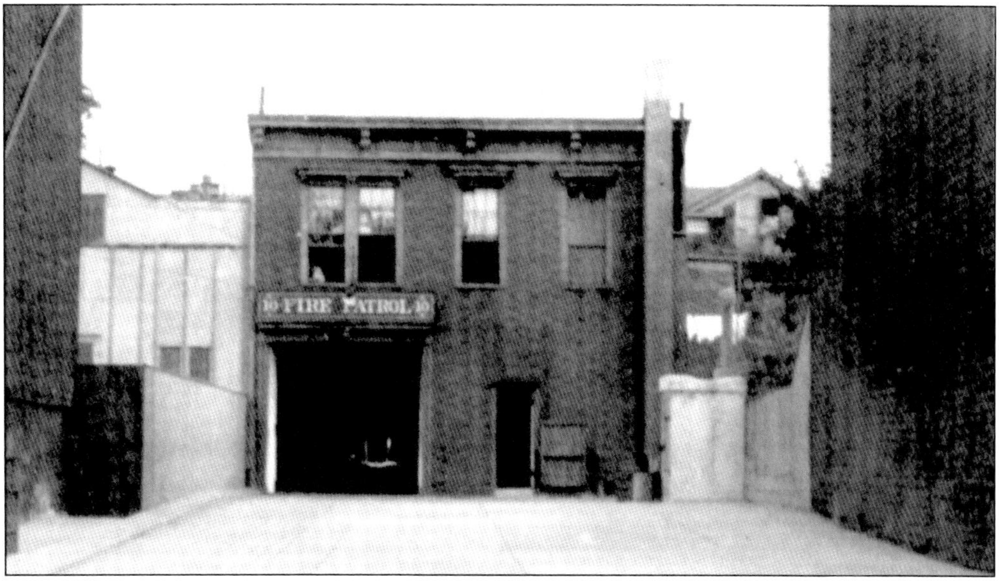

PATROL 10. Organized as Brooklyn Salvage Corps Station No. 3 on December 31, 1905, this would be the last of the patrol units to be formed in the city. Originally located on Brooklyn Avenue near Herkimer Street, they moved to 516 Herkimer Street on January 31, 1906. The Herkimer Street location was the only frame structure to serve as a patrol house and may well have been a converted building. BFISC Station No. 3 was renumbered Patrol 10 on January 14, 1911, and held that numbering until it was closed on May 1, 1954. (NYFP.)

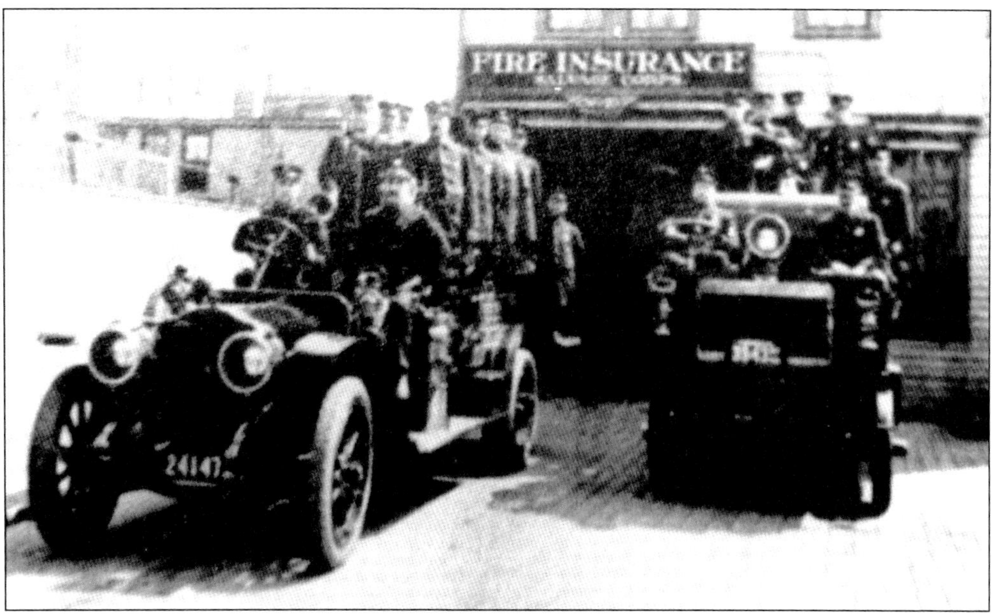

HERKIMER STREET HOUSE. For 48 years, the patrol operated from these cramped, deteriorating quarters on Herkimer Street. During this time, no improvements had been made to the building other than a new roof and asbestos shingles. By 1950, the NYBFU had drawn-up plans for a new building and were searching for a suitable piece of property. This plan never materialized and Patrol 10 became a memory in 1954. The building remained well into the 1980s. (TER.)

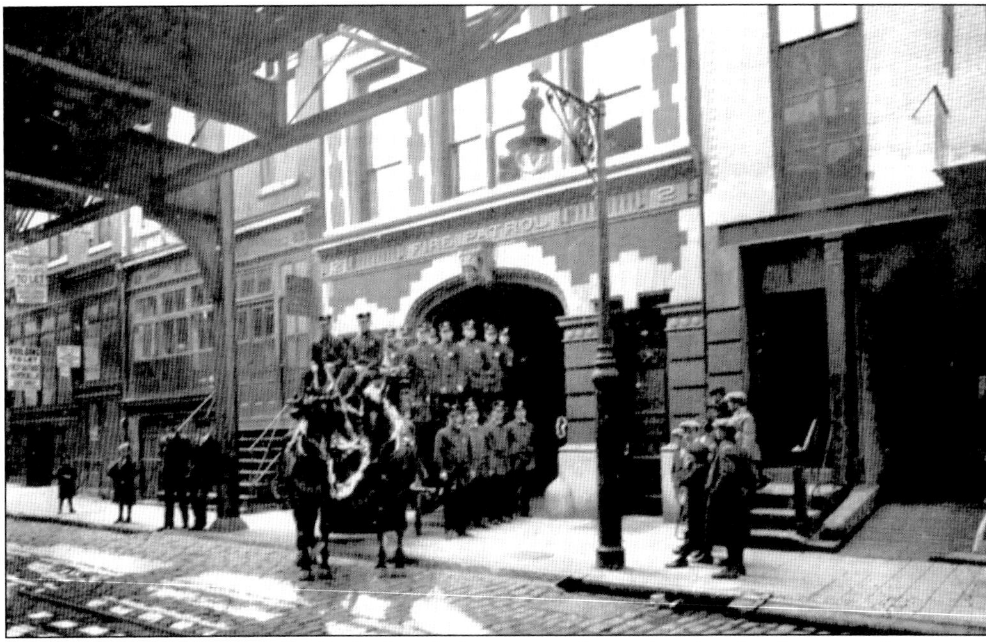

FRONT OF PATROL 2, 1970. Both details have turned out this day for the annual company photograph. Above the men is the old Sixth Annual Elevated Railroad, which passed by West 3rd Street well into the mid-20th century. This area is today the heart of Greenwich Village. (NYFP.)

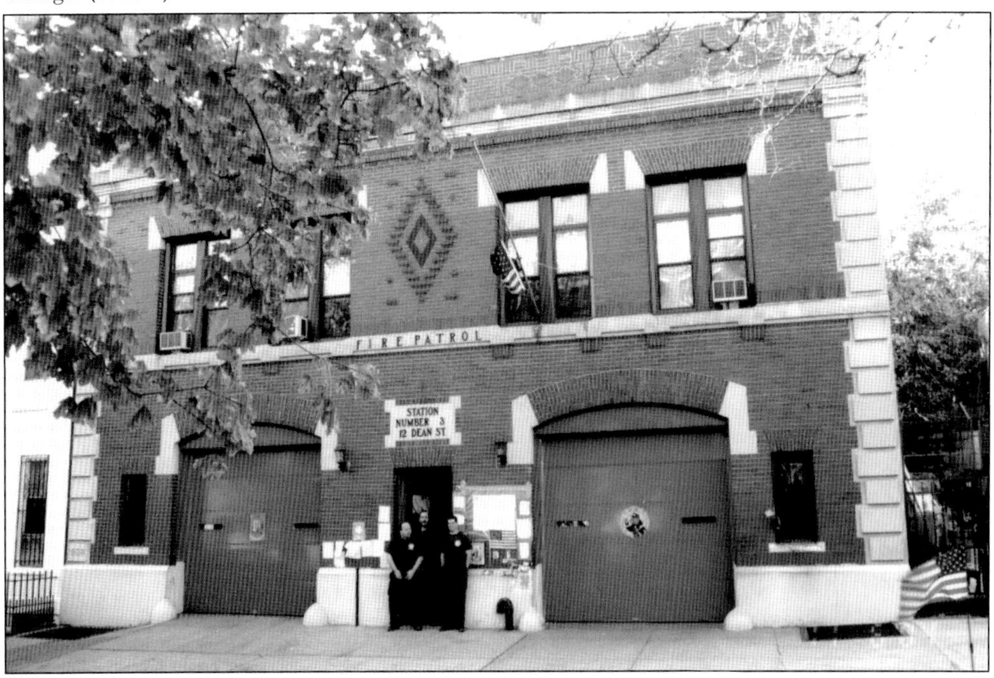

PATROL 3 TODAY. The old quarters of BFISC was merged into the New York Fire Patrol in 1913. Few changes have been made over the years to the structure, although the housewatch man is no longer responsible for firing up the coal furnace and taking the ashcan out to the curb. This building has been well maintained over the years and will be 100 years old in 2008.

Four

Rigs

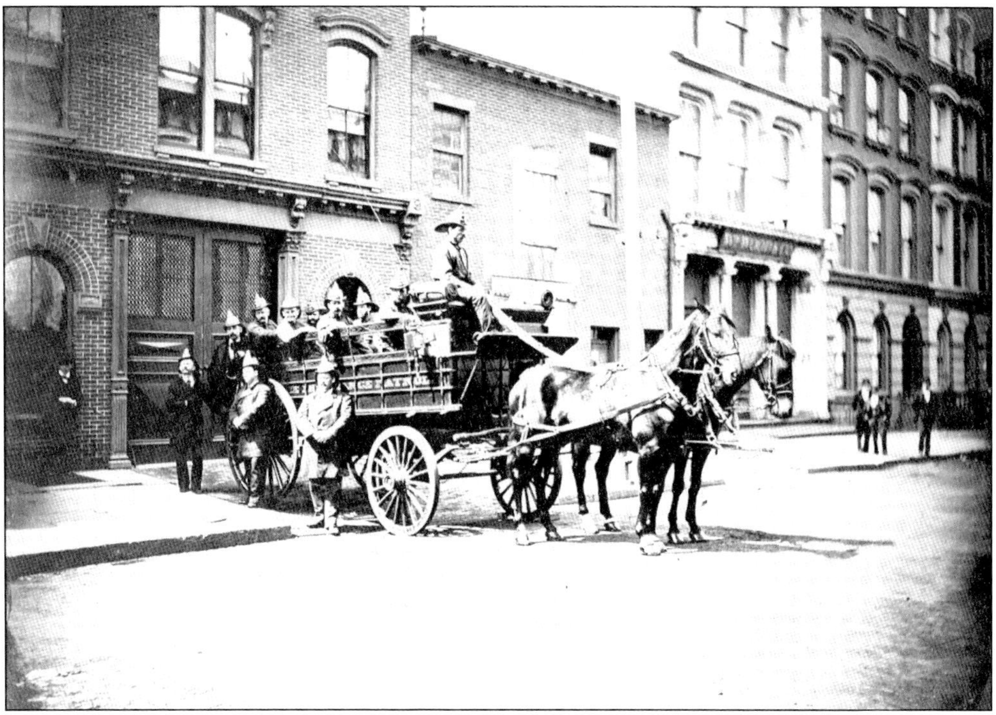

Great Jones Street, 1874. This patrol wagon is of the first type which was universally built for the use of the insurance patrol starting about 1865. It was basically a simple delivery wagon modified to carry the equipment of the patrol. It was the first wagon that the men could ride in and carried a complement of about 40 salvage covers, pry bars, sawdust, lanterns, and other miscellaneous equipment. (FDNY.)

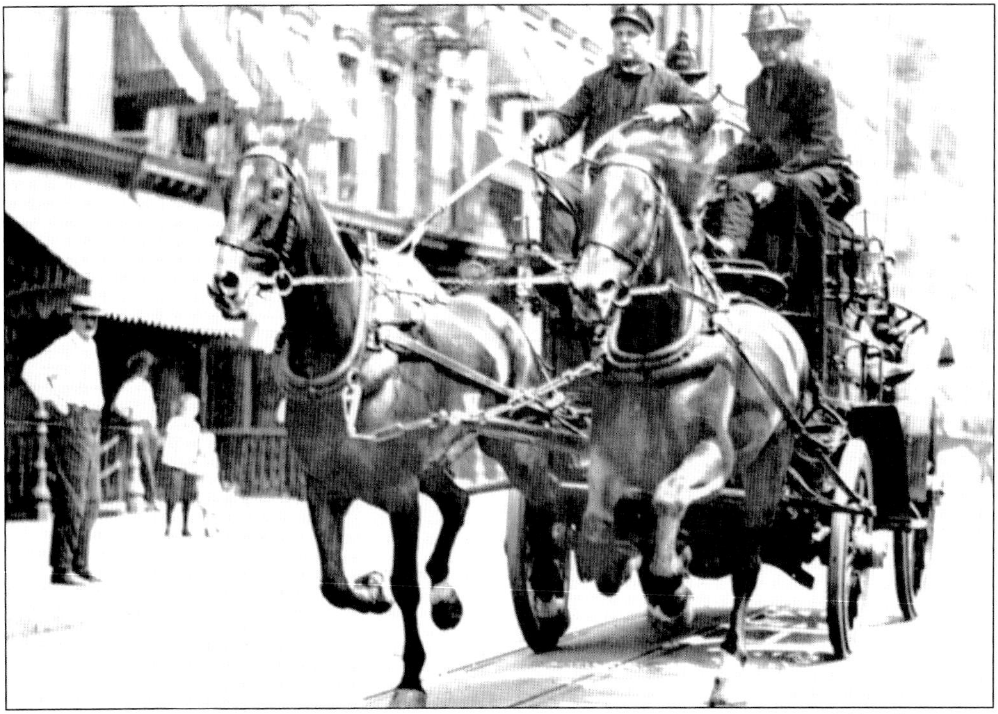

RESPONDING, C. 1902. From 1865 to about 1909, the site of two large horses in full gallop was a sure sign that the patrol was coming to the rescue. During the late 19th century, the patrol and the salvage corps stabled over 50 horses and employed their own blacksmith to care for the fire horses' needs. (FDNY.)

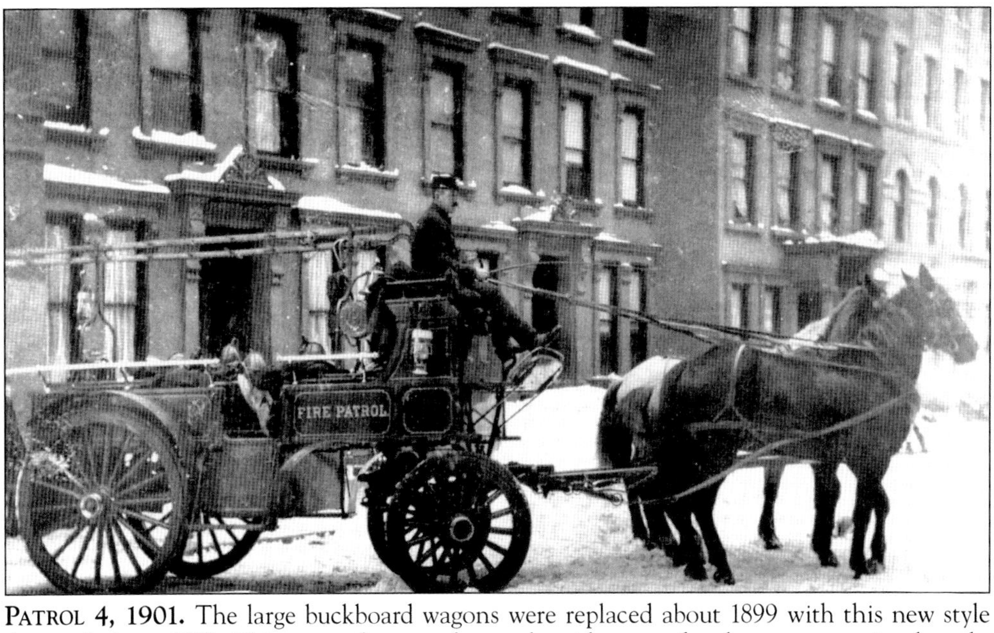

PATROL 4, 1901. The large buckboard wagons were replaced about 1899 with this new style designed about 1889. This type of wagon featured a side entry for the men, running boards, overhead scaling ladder, and improved compartments for tools and equipment. This type of wagon was in use until the time of motorization with no major alterations. (NYFP.)

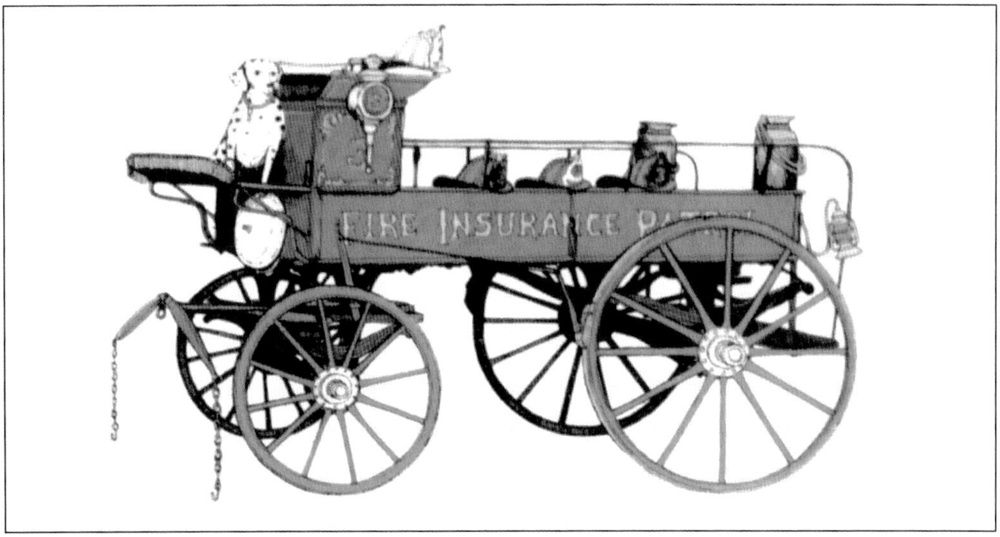

FIRE PATROL WAGON, 1880. The patrol operated a variety of salvage wagons before motorization. The earliest pie wagon gave way to the larger "railroad depot" type wagon. Oddly in 1856, the patrol purchased a two wheel wagon for extra equipment and placed it in service with the lower patrol. This light duty wagon could be pulled quickly by the men, for it would not be until 1866 when horses were first used. (TER.)

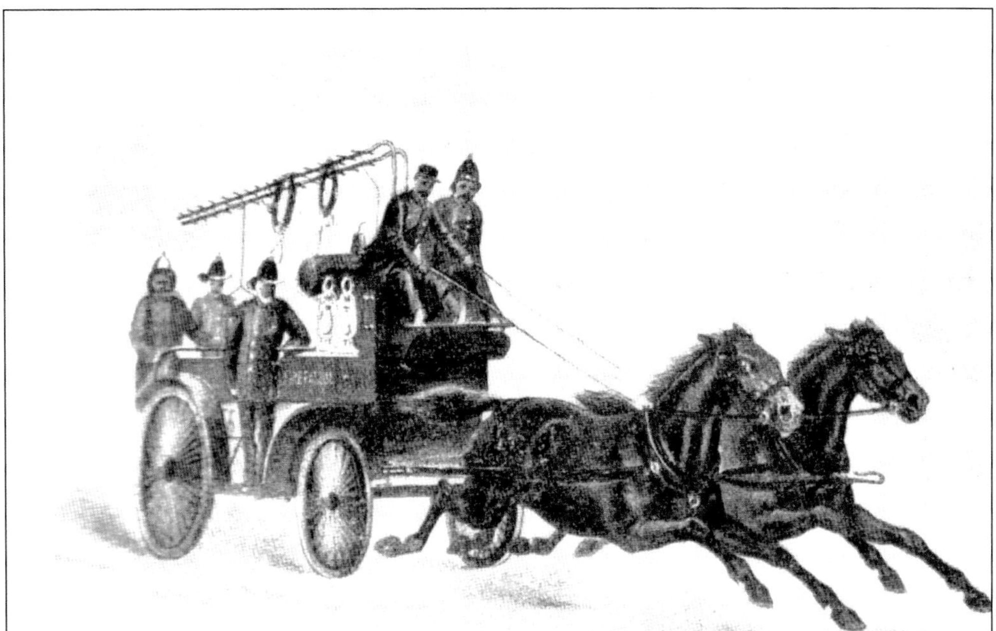

DASHING TO THE SCENE. This charcoal sketch captures the speed and the glamour of the old days. Adults, children, and storekeepers would drop everything to stand at the curb and watch the fire laddies go by.

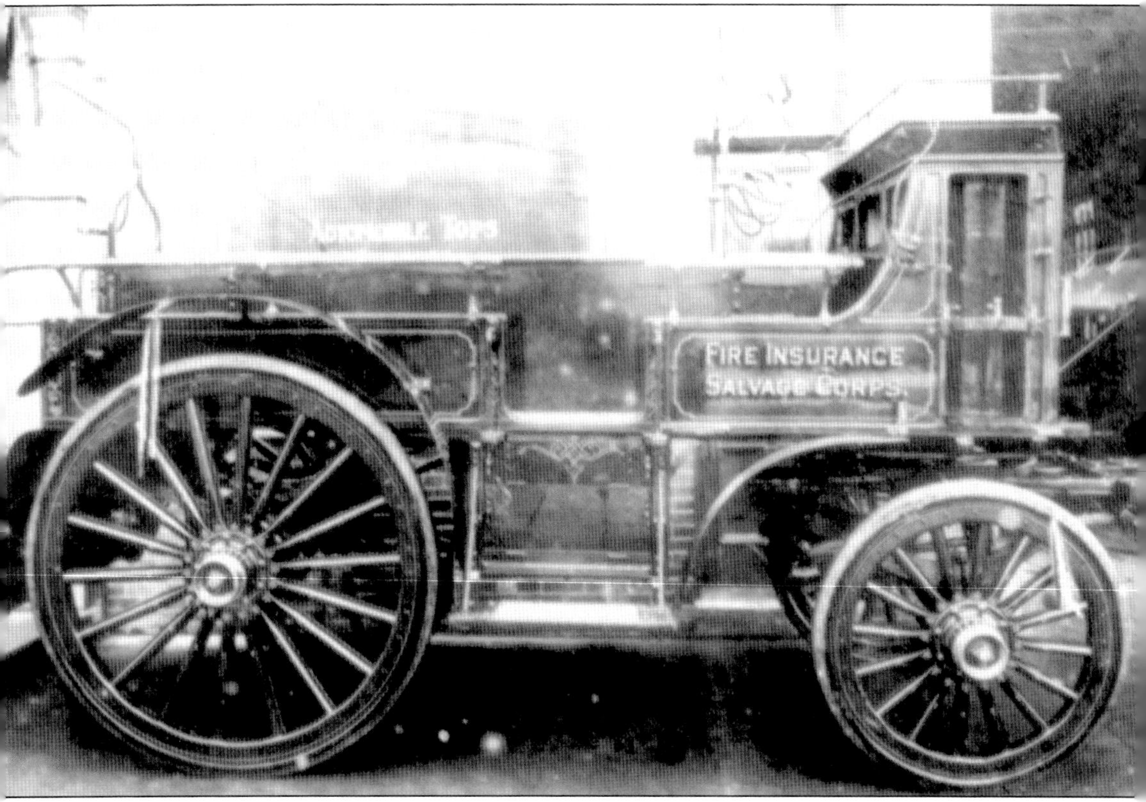

BFISC Salvage Wagon, 1905. All operations of Brooklyn's Fire Insurance Salvage Corps were identical to that of the NYFP. After using a small patrol wagon built locally in Brooklyn, the salvage corps adopted this style wagon, which was of a similar design than that of the city wagons. (JK.)

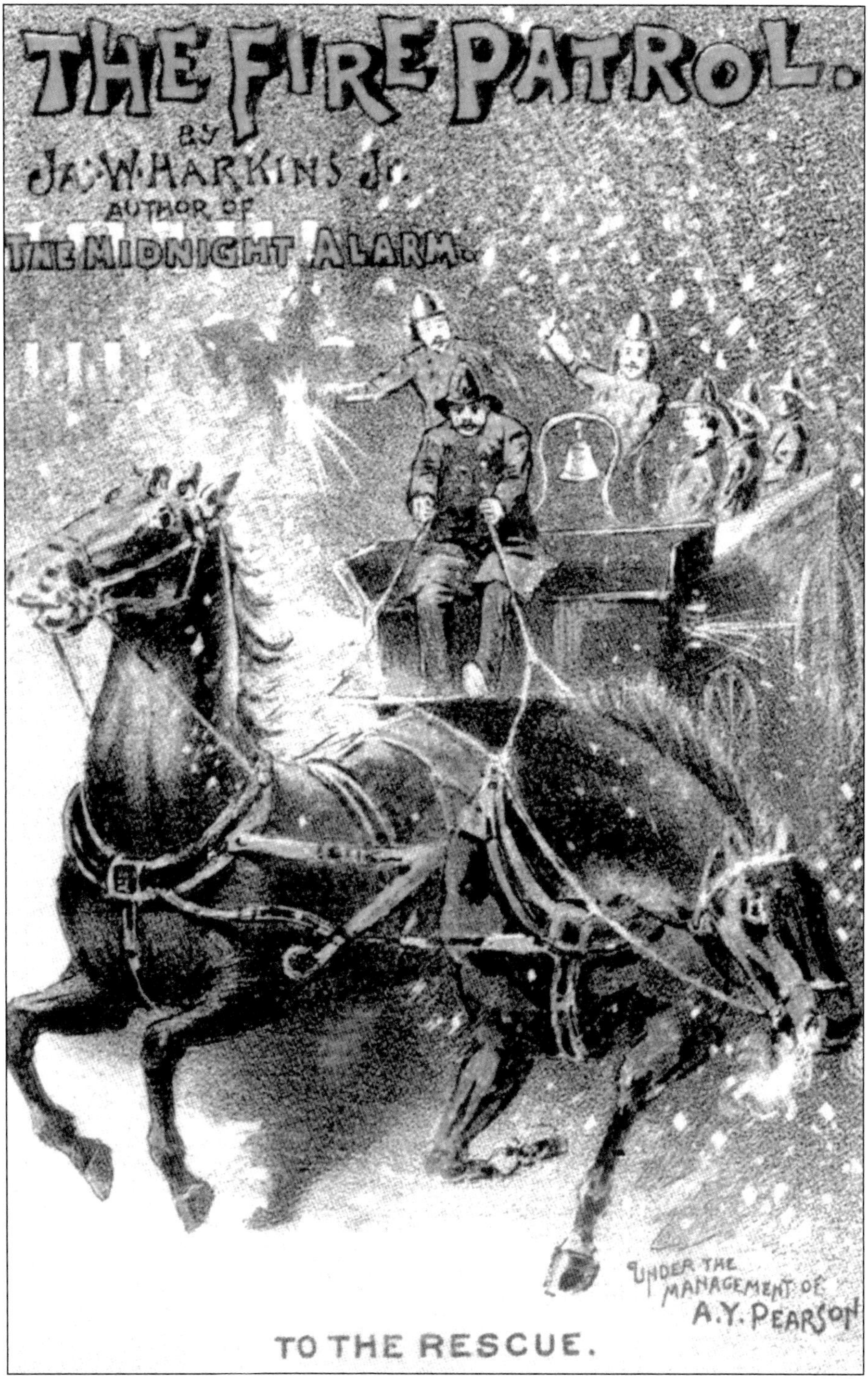

FIRE PATROL PROGRAM, 1900. Theatrical performances, highlighting the dangerous exploits of the fire patrol's work, appeared on stage several times in the early 1900s. Here the crew of patrolmen are making their way through a fierce snowstorm; of course they wanted good seats for the performance, and who could blame them? (LW.)

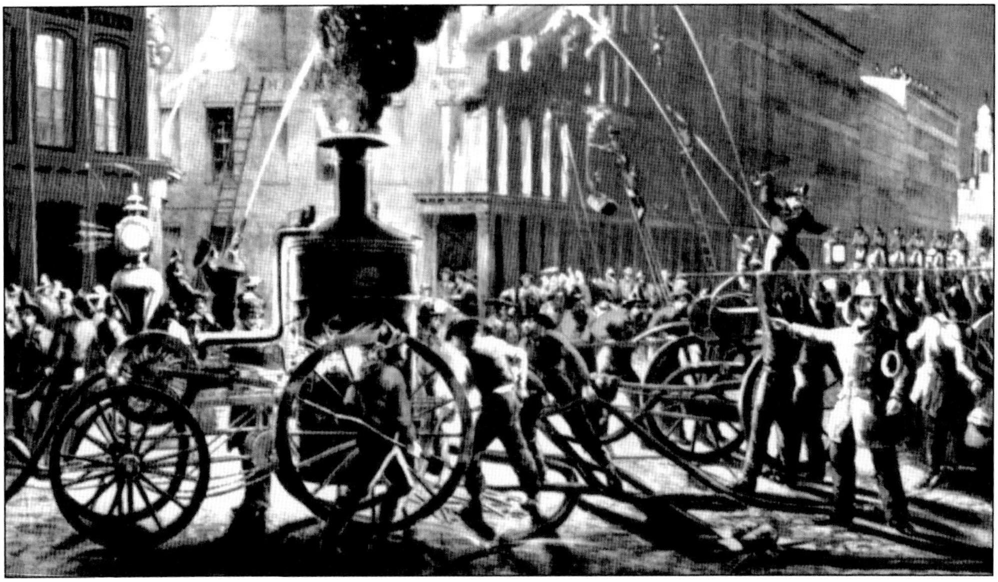

The Steam Engine "Elephant," 1859. On November 15, 1865, the board purchased the 5,600-pound engine from the city. Built by the firm of Lee and Larned, it had been presented by the Insurance Companies to Manhattan Engine Company No. 8 in 1859. Flooded basements, cellars, and even ship hulls were no match for the Elephant, who retired in 1881 after a long career in New York City. The Elephant was immortalized in the 1861 Currier and Ives Lithograph *The Life of a Fireman*. (TER.)

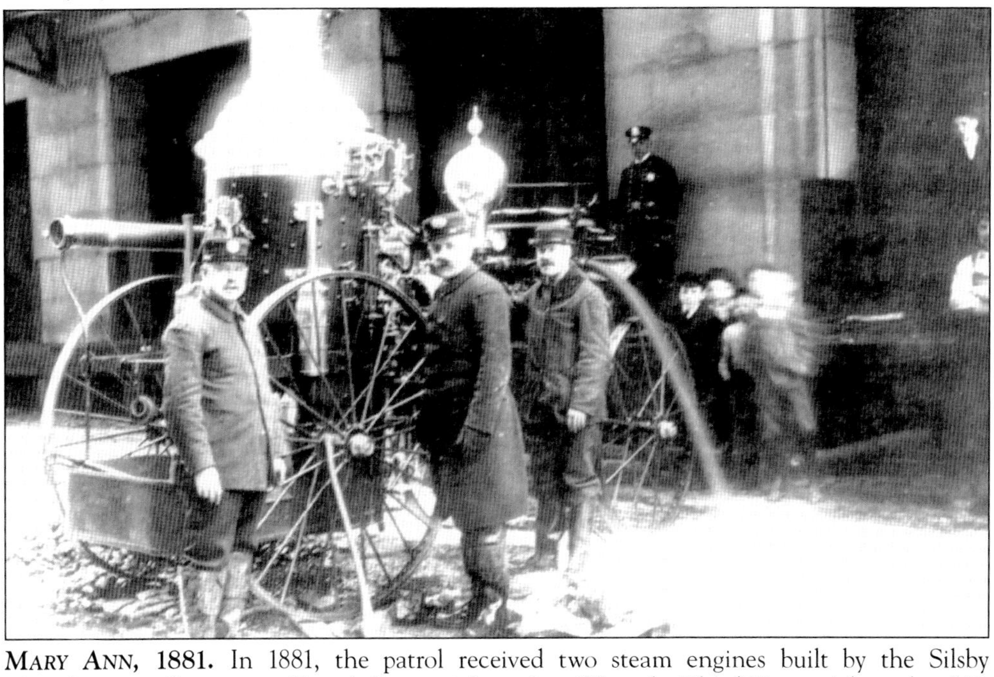

Mary Ann, 1881. In 1881, the patrol received two steam engines built by the Silsby Manufacturing Company, a Fourth Size, serial number 613, and a Third Size, serial number 669. *Mary Ann* was operated by Owen Daily for many years and served in first line status until 1917. It was kept well into the 1930s at Patrol 6, finally giving up its life, like so many other steamers, for a World War II scrap drive. (NYFP.)

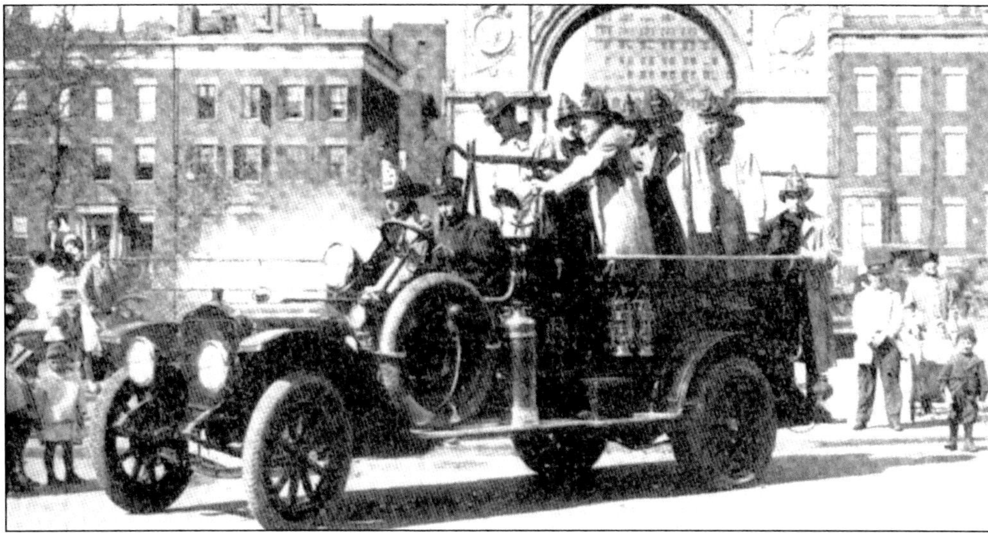

WE'RE ON OUR WAY, MEN. The patrol is riding with a full crew in the detail this day, and they are going to need it. A small fire has broken out in an old loft building on Broadway and sprinklers are flooding the priceless reams of silk. It is going to be a big cover throw, the captain better send out the alarm for two more patrol wagons to respond.

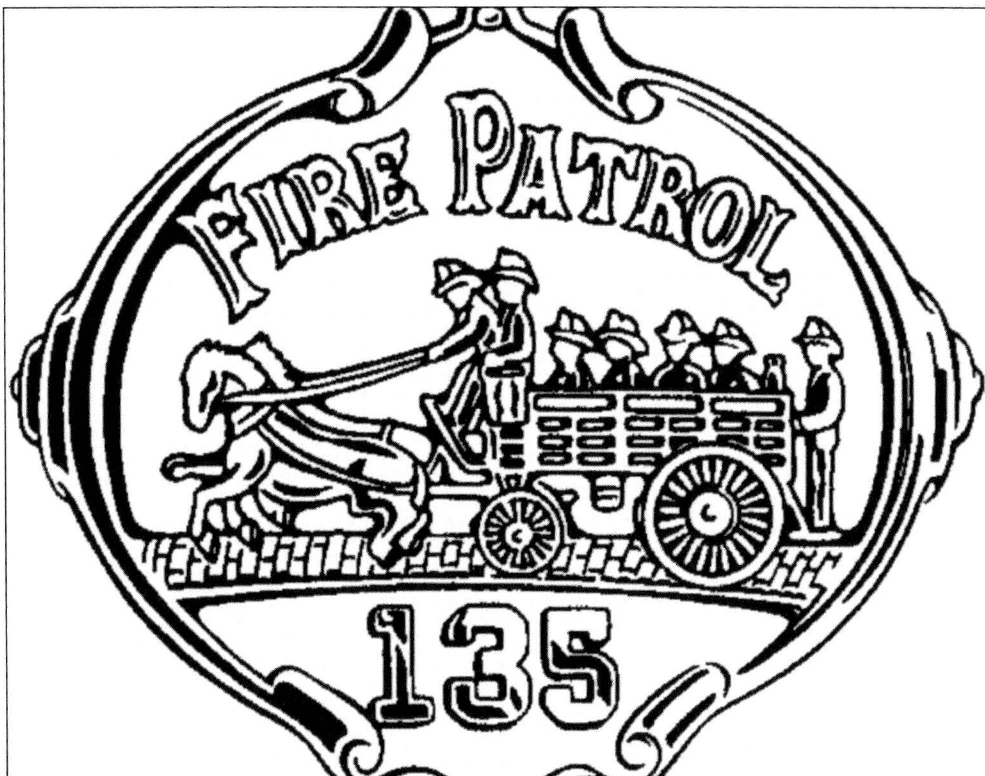

FIRE PATROL BADGE. The old badge of the patrol was first introduced in 1890 and continued until about 1939. It featured a full complement of patrolman, officer, and a chauffeur rushing to a blaze on a cobblestone street. (SG.)

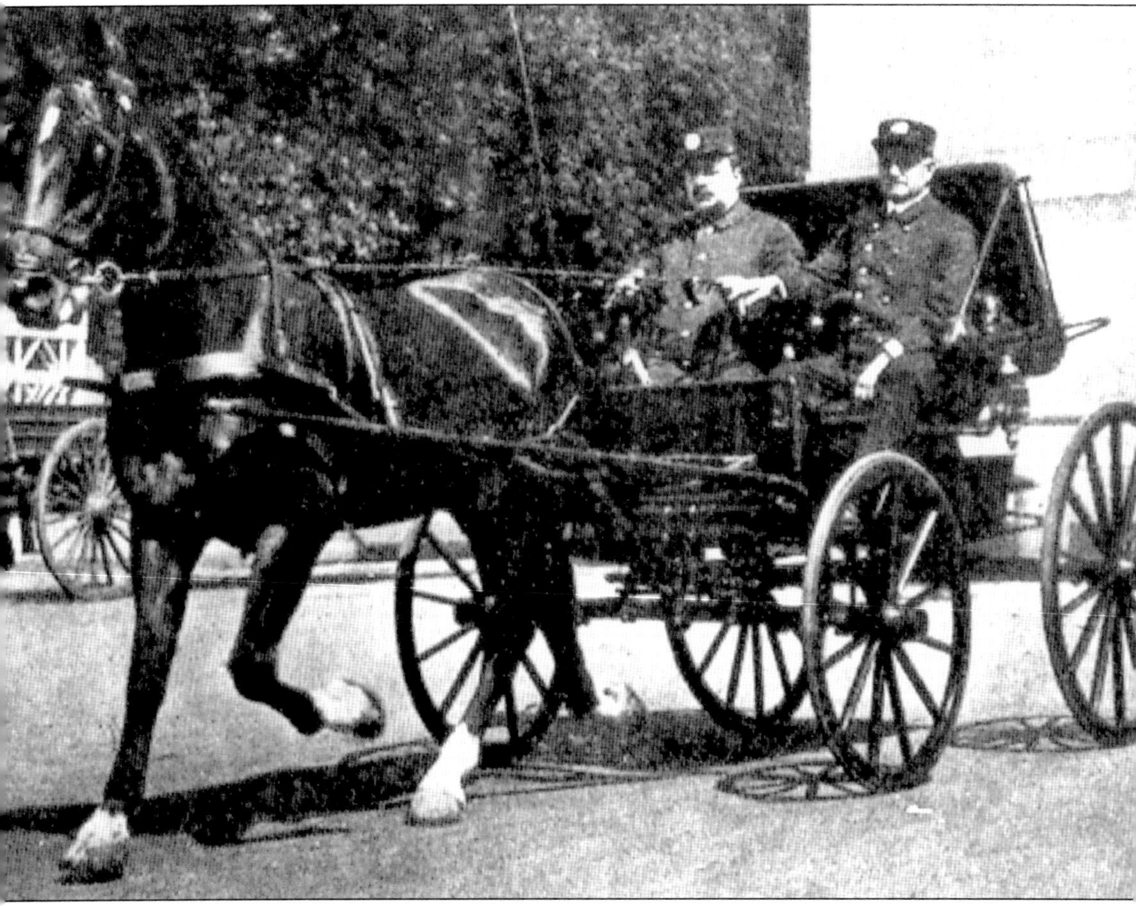

SUPERINTENDENT'S BUGGY, 1900. Frederick Groves was appointed to fill the position of superintendent of the patrol in 1899. He is riding in a single horse buggy, which was used until 1910, when a Knox automobile was purchased.

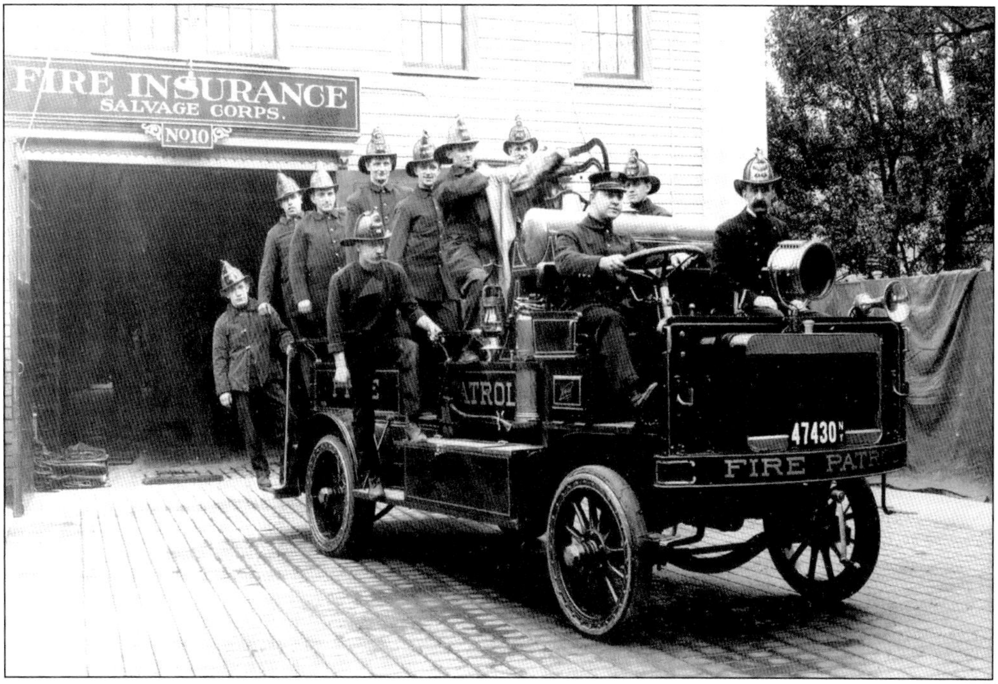

MOTORIZATION ARRIVES ON HERKIMER STREET. This 1911 chassis was purchased from the Knox Motor Company located in Springfield, Massachusetts. Several of these rigs were purchased for the patrol, at which time the older horse drawn bodies were mounted on the new Knox Chassis. Powered by a 40 horsepower water cooled engine, the Knox was capable of speeds up to 28 miles per hour, increasing response times and once again saving property that would otherwise be ravaged by fire and water. (TER.)

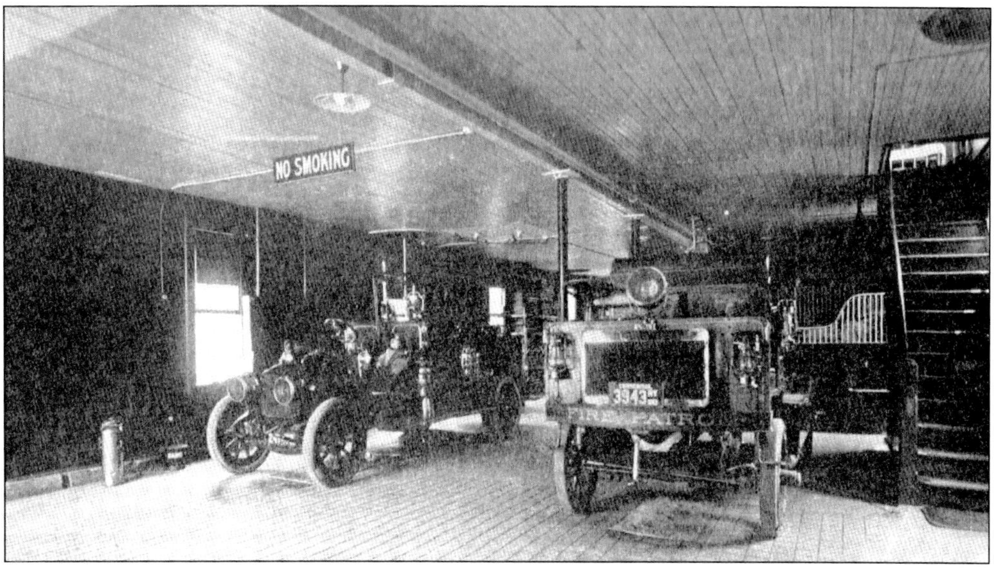

INTERIOR PATROL 10, 1917. Here the two sections of Patrol 10 wait for another Brooklyn alarm. In the background is a c. 1915 White truck with a patrol body built by Peter Pirsch and Sons located in Kenosha, Wisconsin. Next to the White is another 1911 Knox (right). Note the Knox has no headlights, only a large searchlight and lanterns. (NYFP.)

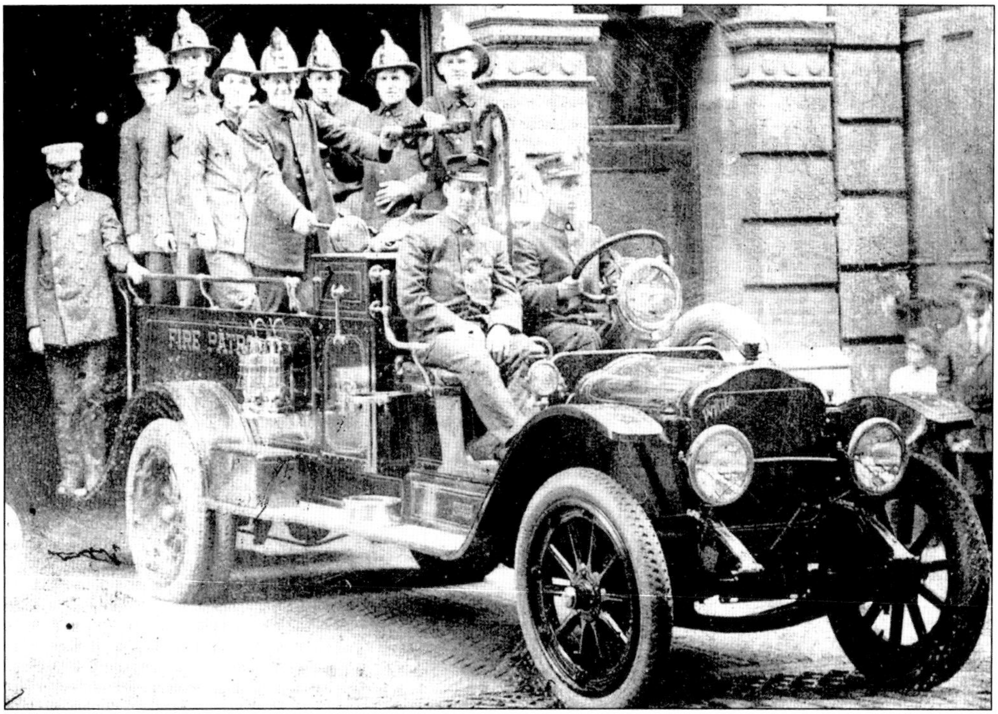

PATROL 2 TURNS OUT. Pictured here is the infamous White motor truck of Patrol 2. On February 9, 1916, the detail was responding to a fire at 221 East 19th Street, when the truck crashed into a streetcar. The men were thrown from the truck and Capt. Peter Cornwall was crushed between the streetcar and the truck. Surgeons amputated his leg at a nearby drugstore, and he died the next day after 32 years with the patrol. (FDNY.)

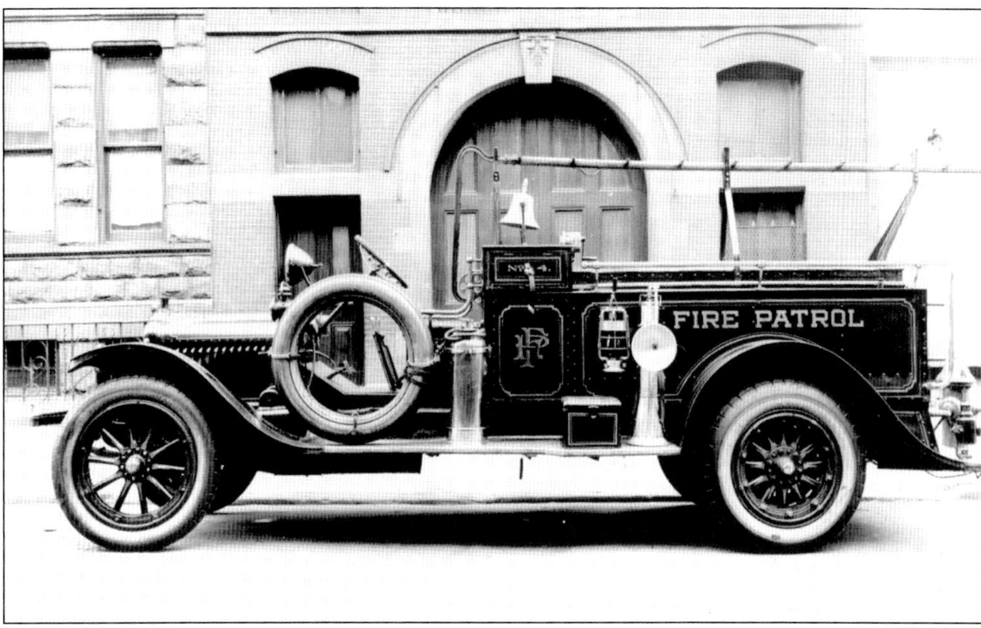

PATROL 4, 1921. This *c.* 1919 White of Patrol 4 is seen here shortly after its delivery to the East 90th Street firehouse. These trucks featured wooden spoke wheels and hard rubber tires. (NYFP.)

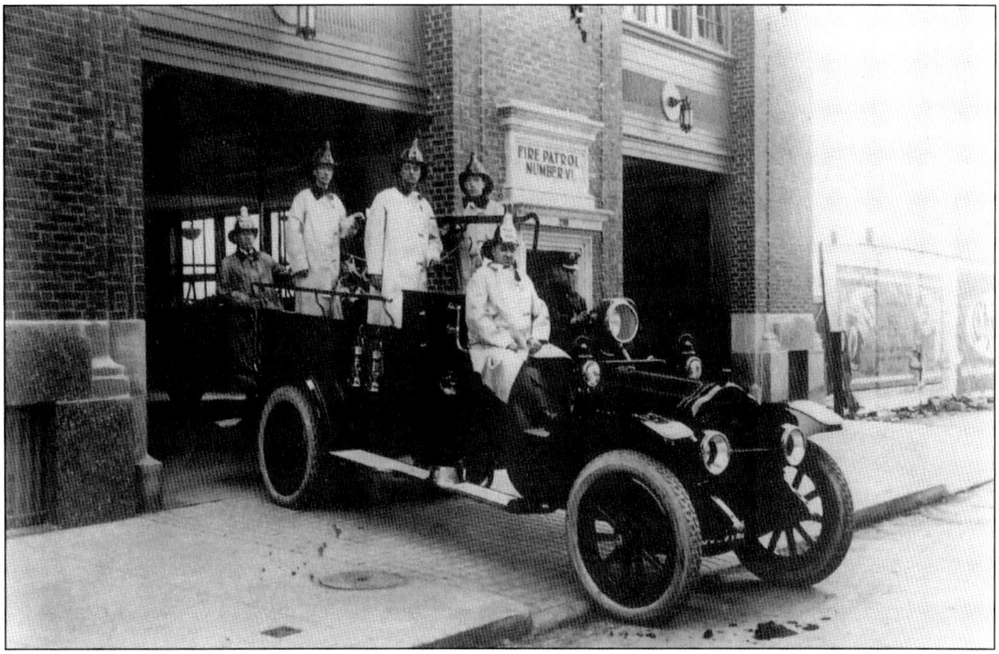

BOYS FROM THE BRONX, 1926. Pulling out of the firehouse at 278 East 156th Street is the working detail of Patrol 6. The men are wearing the old white rubber coats which were in vogue for over 60 years. Unfortunately, the vulcanizing process used to manufacture the coats caused them to eventually become brittle and turn an ugly shade of yellow. By 1930, they had disappeared. (NYFP.)

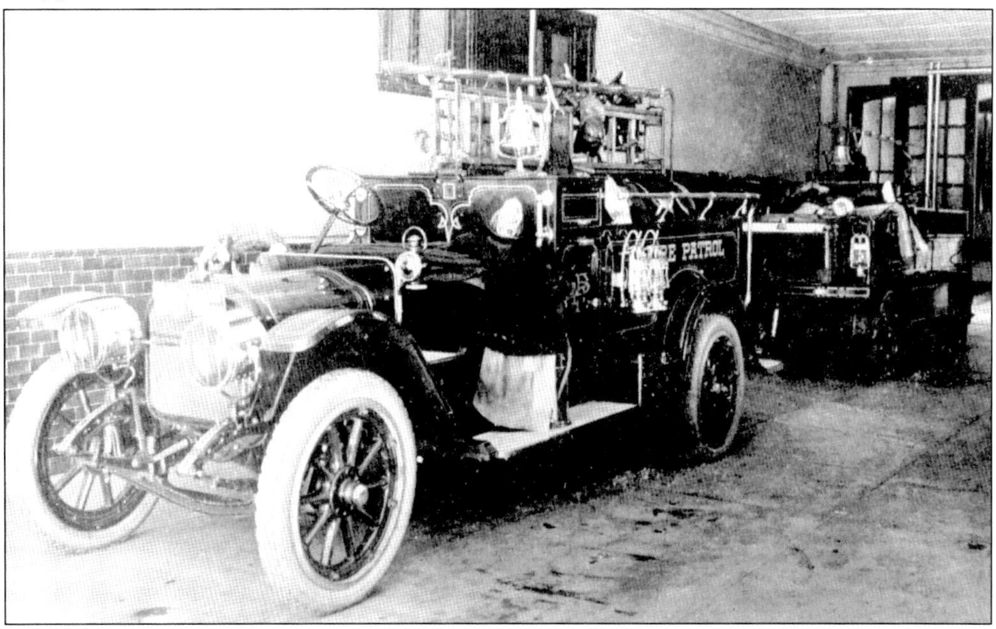

PATROL 5, 1917. Approximately 17 White/Pirsches were placed in service before 1920. Of them, Patrol 5 seems to be the only one equipped with scaling ladders and an extension ladder. This may be the earliest of the Whites, as evidenced by the right hand drive, very early style headlights, and the small lanterns mounted on each cowl. (NYFP.)

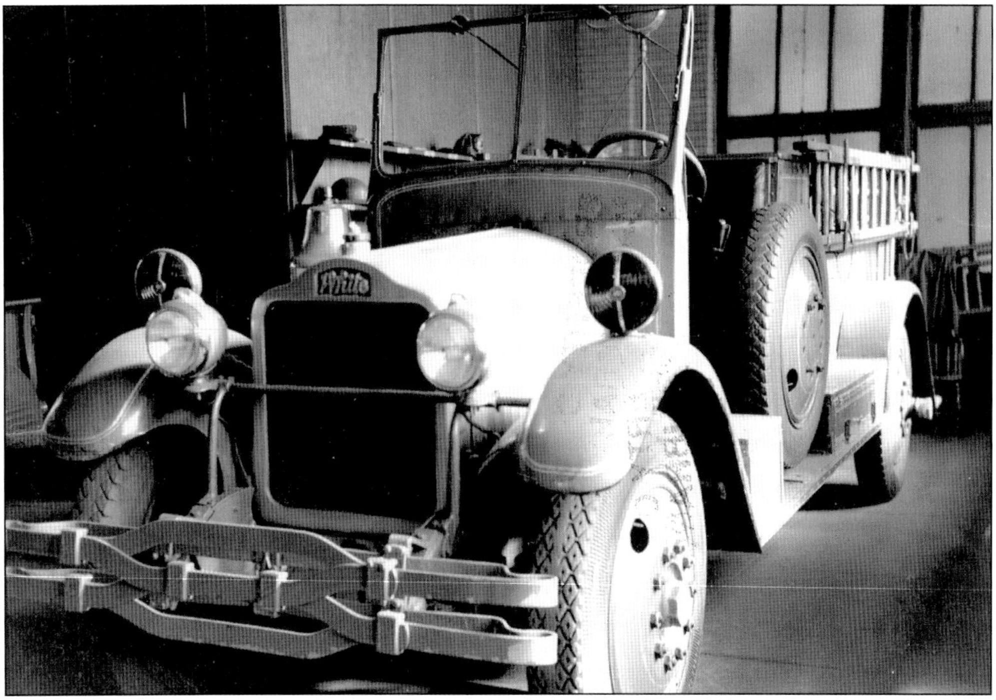

WHITE PATROL TRUCK, 1940. At some point in this early White's life, it was sent to the shops and completely modernized. Additions to this rig included a heavy duty bumper, pneumatic tires, large fender mounted flashers, a windshield, and an extension ladder. It appears this truck was being used as a reserve truck located at Patrol 5 when this photograph was taken. (FDNY.)

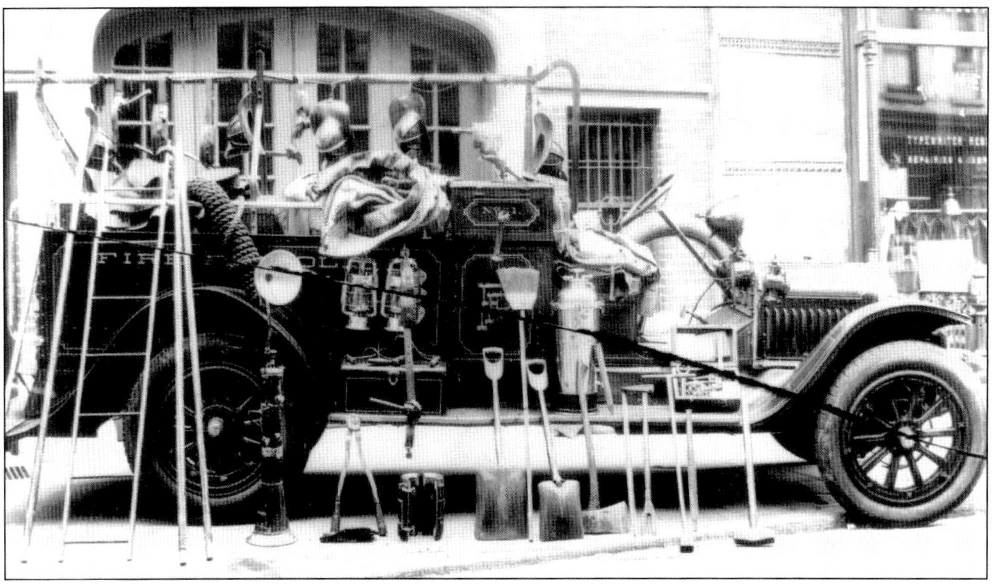

PATROL 1 EQUIPMENT, 1919. Despite operating for over 170 years, the basic tools of the fire patrolman's trade have changed little in that time. This early 20th century photograph shows brooms, axes, pry bars, acetylene powered light, bolt cutters, fire extinguishers, rope, scaling ladders, and of course salvage covers. During this time, the patrol had an inventory of 4,000 of the 18 feet by 12 feet stock covers, and 500 roof covers located in a total of 10 firehouses. (JK.)

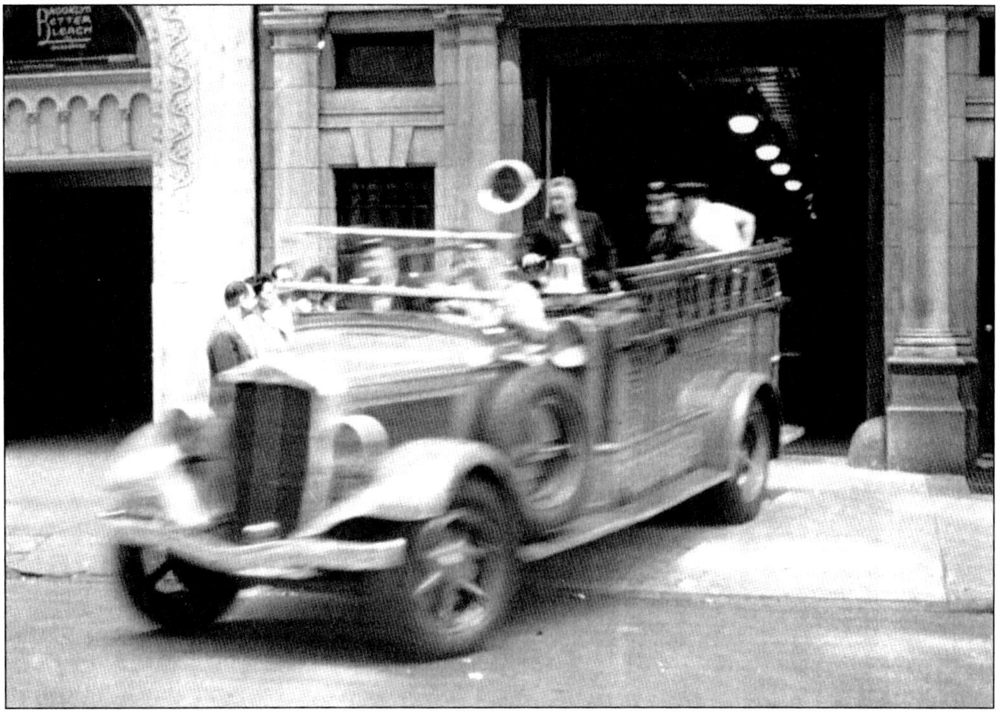

PATROL 1'S DETAIL, 1937. In 1934 and 1935 four new White trucks were purchased to begin the replacement of the aging fleet of earlier Whites. These new rigs were delivered with open cabs, two sets of extension ladders and windshields, and were built on newly styled chassis. They contained six cylinder 105 horsepower engines which had much more power than the old Whites. (PW.)

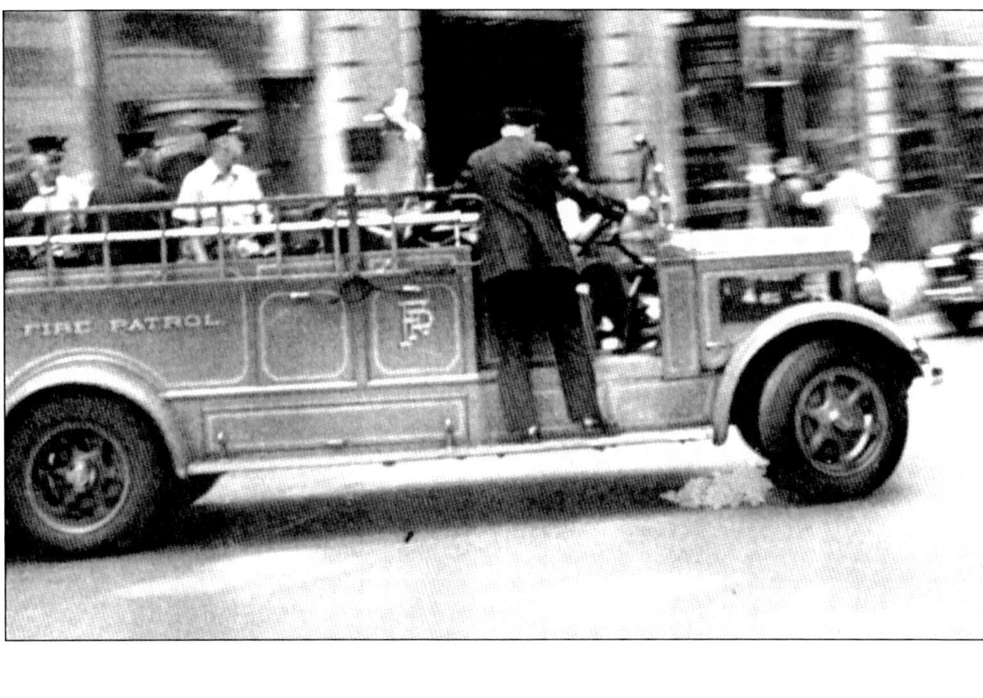

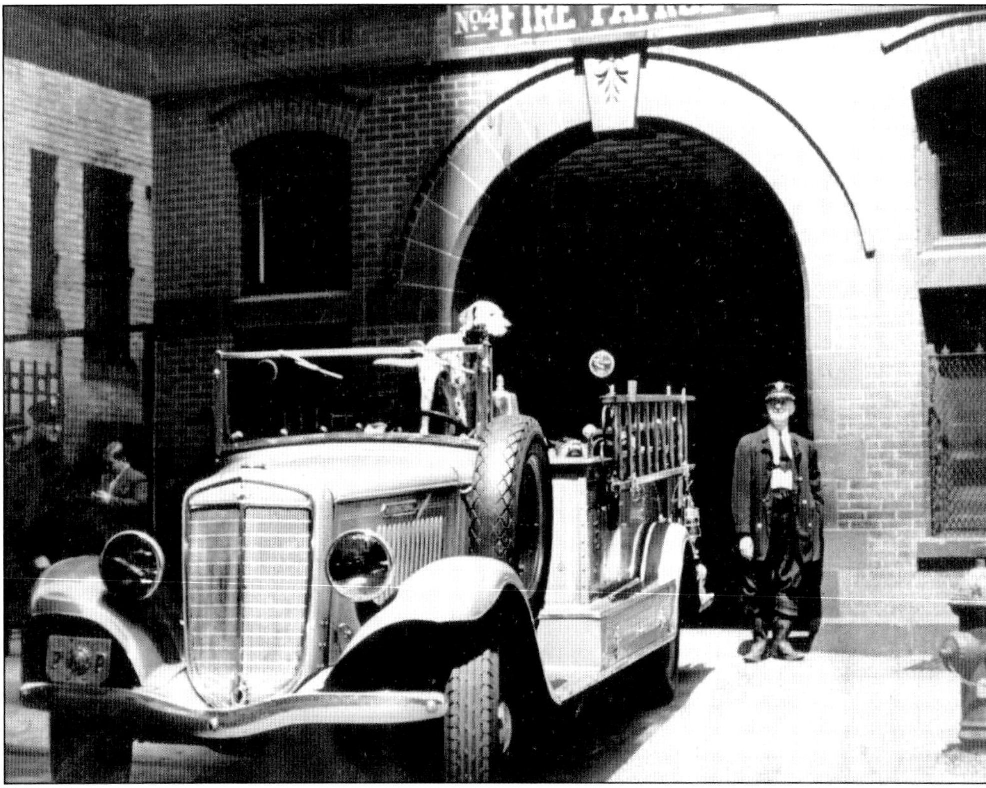

PATROL 4, 1939. The International Harvester Corporation delivered three of these good looking rigs in 1935. These three trucks had bodies which were of a unique type and manufacturer. Here a patrolman and the company's mascot wait around on the apron of the East 90th Street firehouse just before World War II. In 1950, all three of the Internationals were disposed of at auction.

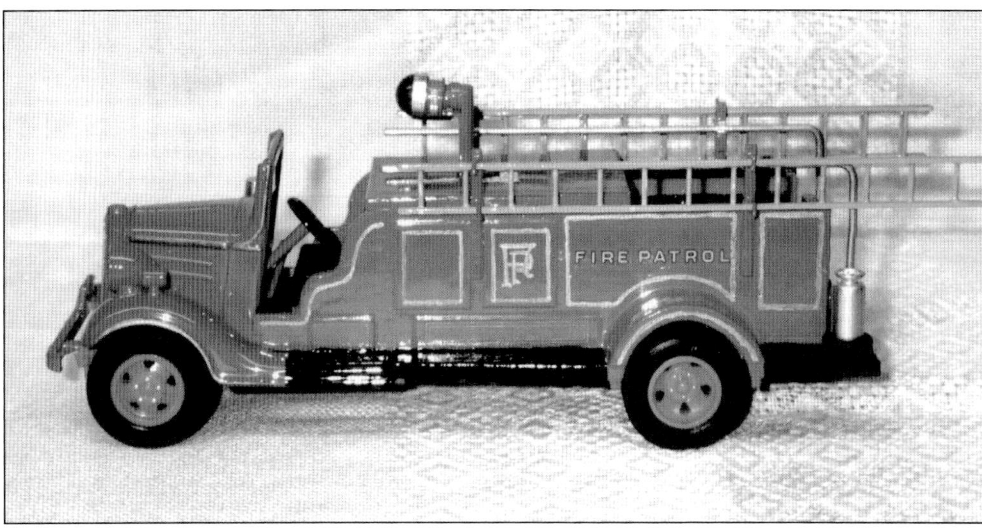

MODEL TRUCK. Built by Paul Wormsley, this scale model patrol truck has all the features of its big brother. It is a replica of an early model White, however it is very similar to the International truck pictured above. (MK.)

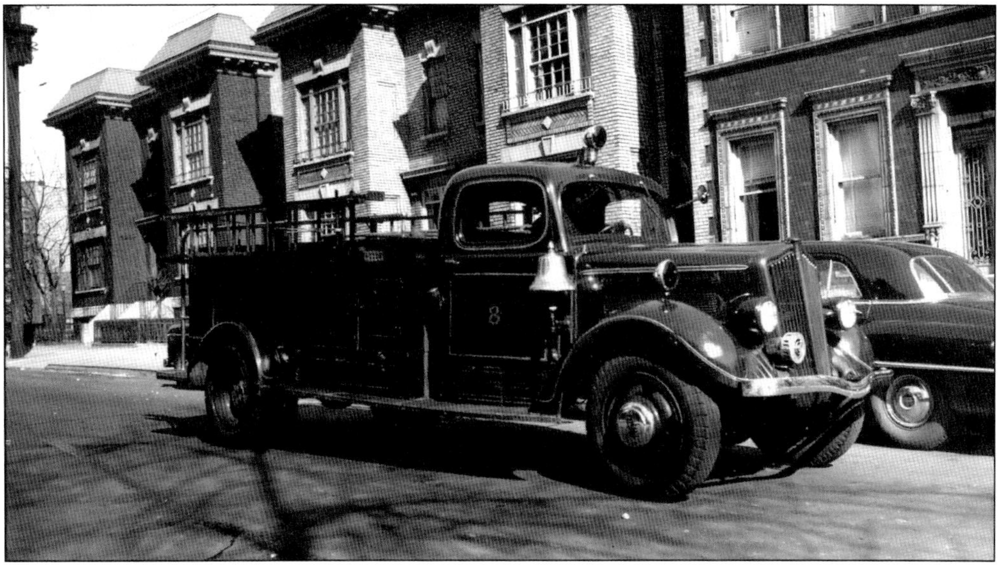

ENCLOSED CAB WHITES. Between 1936 and 1940, all patrol companies received enclosed cab rigs. These new Whites were the first trucks to have fully enclosed cabs for the chauffeur and officer; it would be over 20 years until attempts would be made to enclose the body. These new trucks had redesigned bodies which contained extra storage. By this time, all scaling ladders had been removed and replaced with extension ladders on the officer's side and a roof ladder on the chauffeur's side.

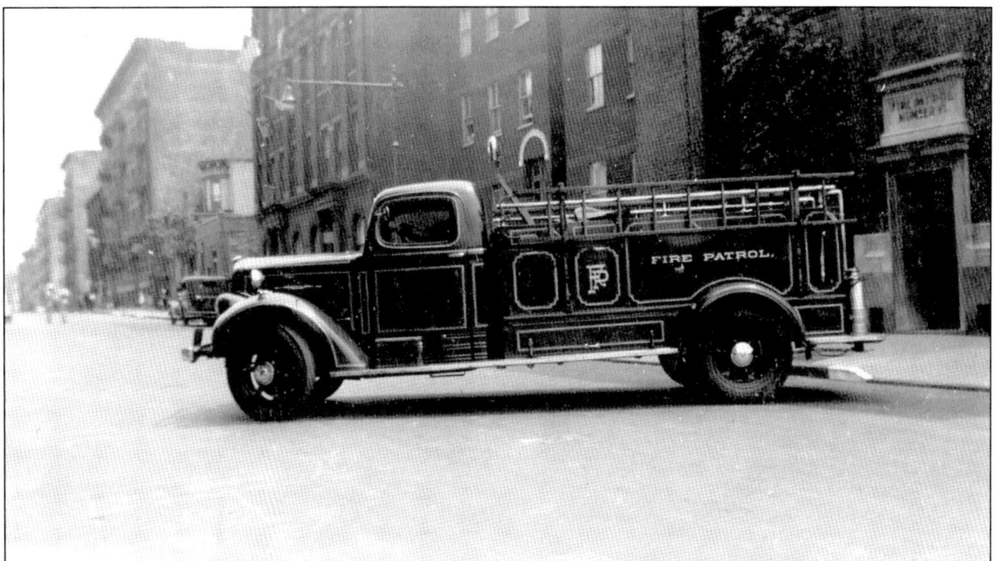

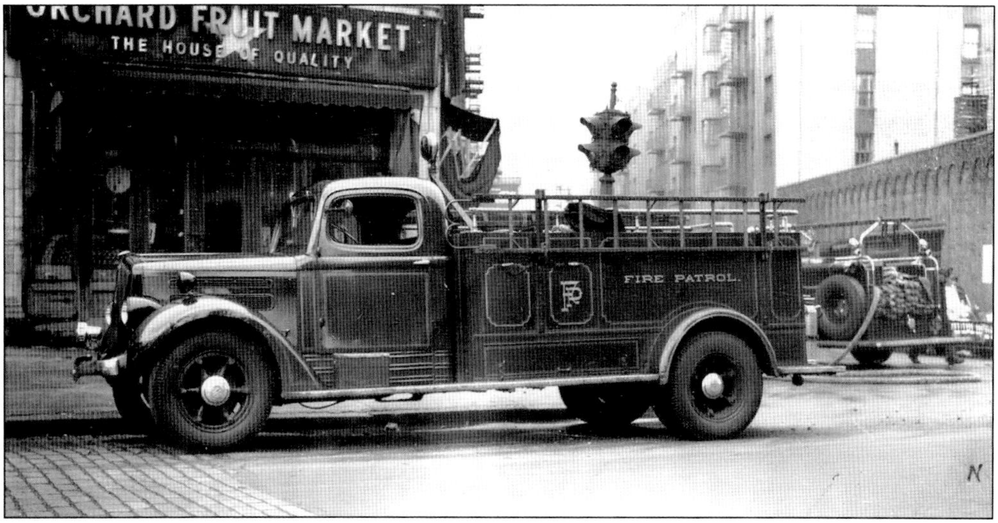

ENCLOSED CAB WHITES. These new rigs were "aerodynamically styled" according to sales literature. The new grille had vertical louvers along with front fenders that swept to the rear. Warning equipment consisted of a bumper mounted mechanical siren, cowl mounted bell, and a large red blinking light post mounted behind the cab. (FDNY.)

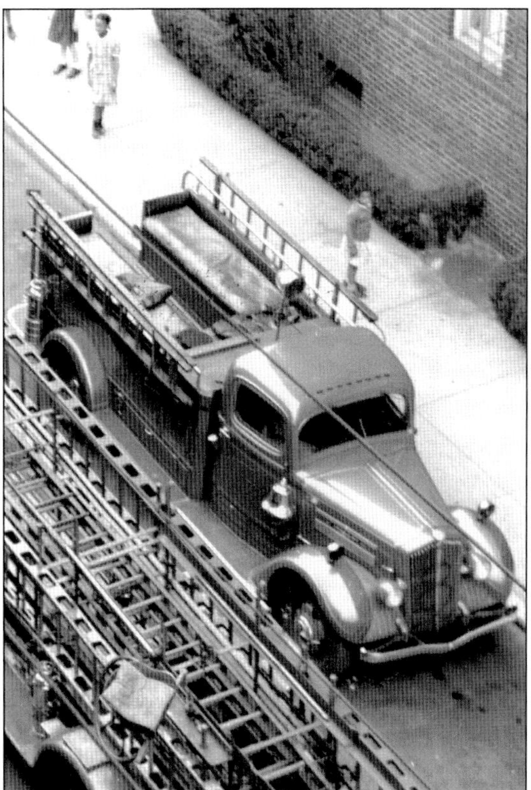

AERIAL VIEW OF A WHITE. Taken during a 1940s job, this photograph allows for a seldom seen view into the rear arrangement of the patrol truck. Behind the cab were four enclosed compartments where sawdust, lights, tools and pumps were carried. The open body contained two large bench seats which were hinged to allow access to the covered storage area. This basic arrangement was in place from 1865 to 1973.

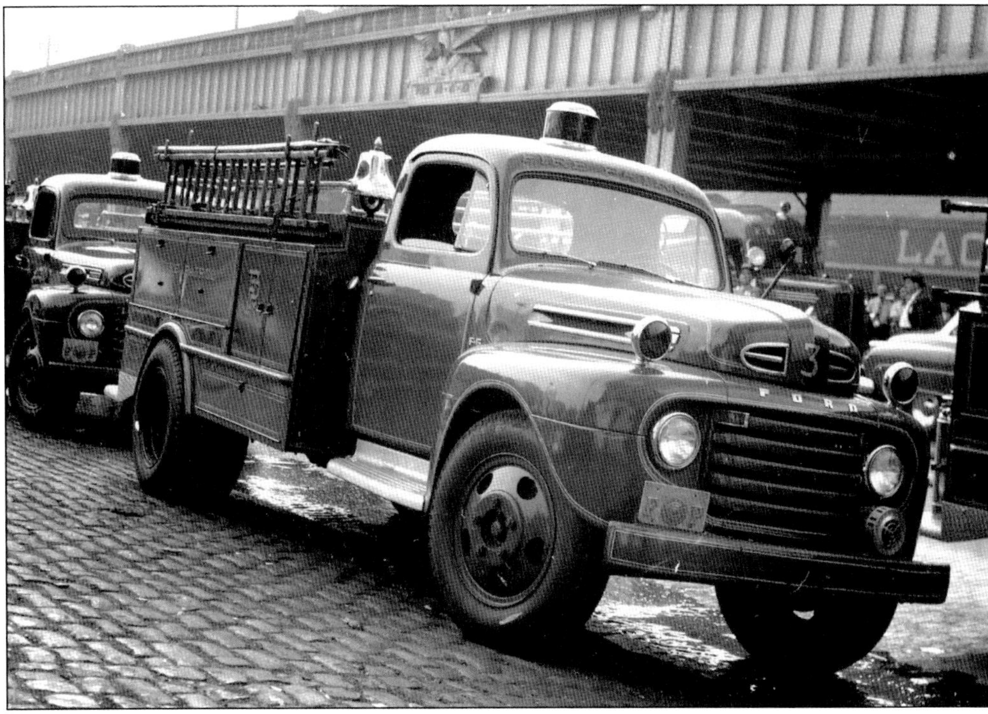

FIRE ON THE WESTSIDE C. 1956. Here Fire Patrol 3's truck sits near the old Westside Elevated Highway, with no less than three other patrol trucks also on the scene. In the background is the 1948 Mack L of Rescue Company No. 1. This was most likely taken at one of the major pier fires during that decade.

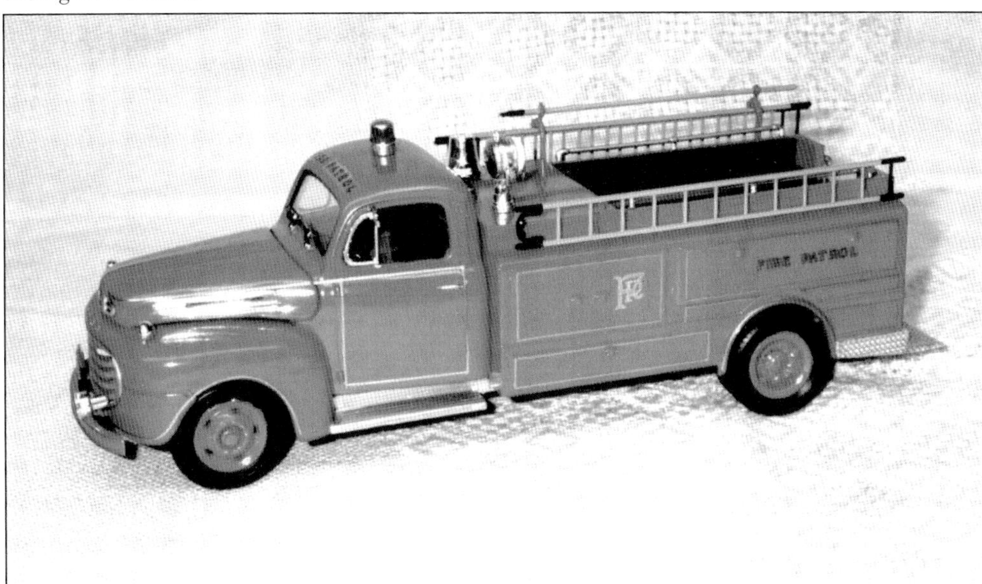

FORD F-5 SCALE MODEL. This outstanding model was scratch built by master fire apparatus miniature builder Paul Wormsley. He has constructed many detailed replicas of American fire apparatus over the past 30 years, and his work is highly sought after. Wormsley himself served with the patrol from 1971 until 1978.

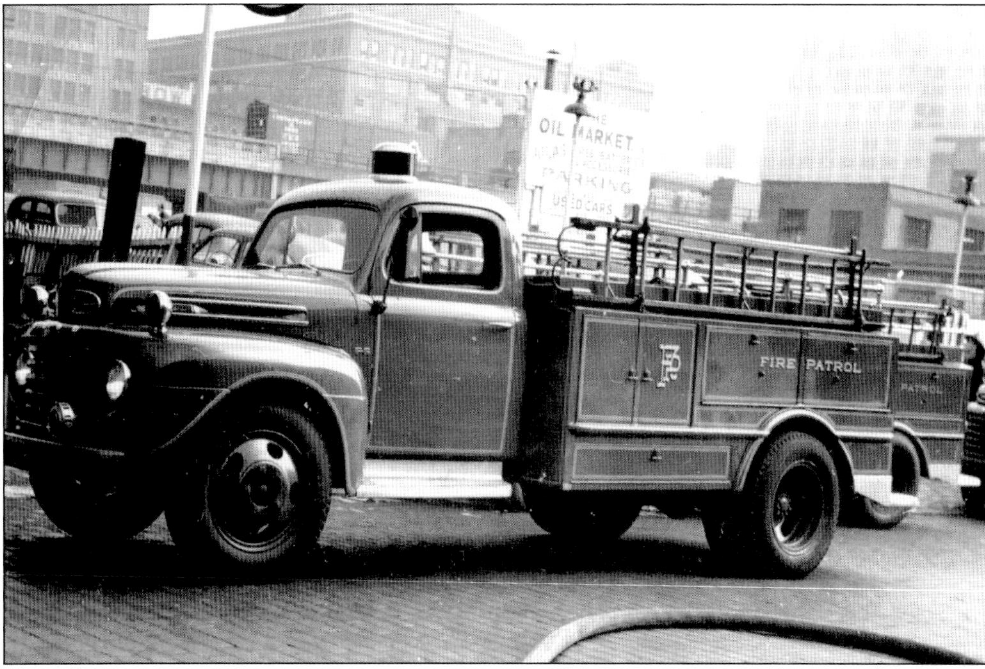

FORD F MODEL. The new Fords were capable of hauling over two tons of equipment and were well known for the speed and good turning radius on the city streets. These new Fords also were the first to be equipped with the new light from the Julian McDermott Corporation in Queens. This new style light contained four flashers which intermittently flashed from front and back and then left to right. It was the forerunner of the modern warning light, and people knew it had to be the patrol when they saw the flash coming.

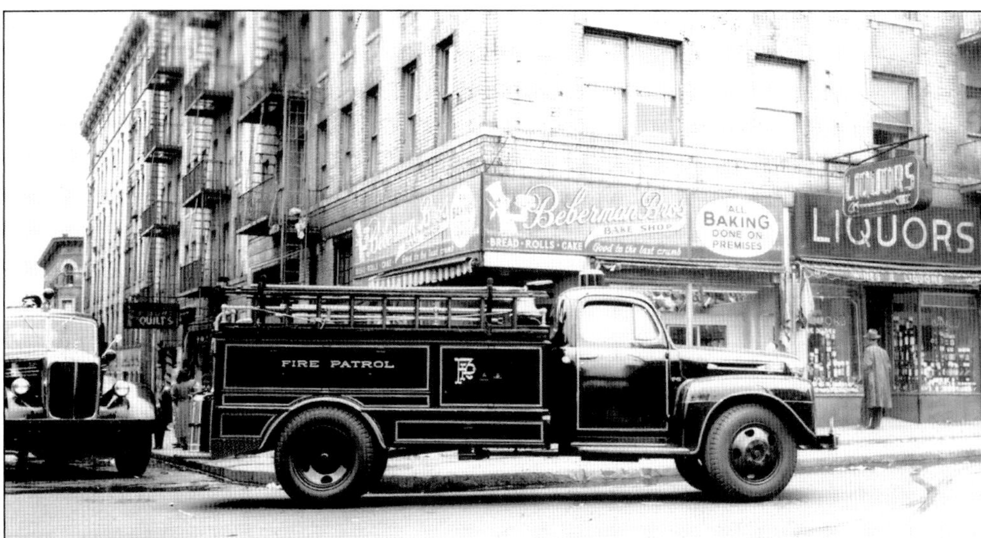

FORD AND MACK. Here an old Ford sits alongside a Mack L model pumper from the fire department, at an unidentified blaze in the mid-1950s. The patrol never owned a Mack, they were preferred by the fire department, and were used for some time with the Philadelphia Fire Insurance Patrol.

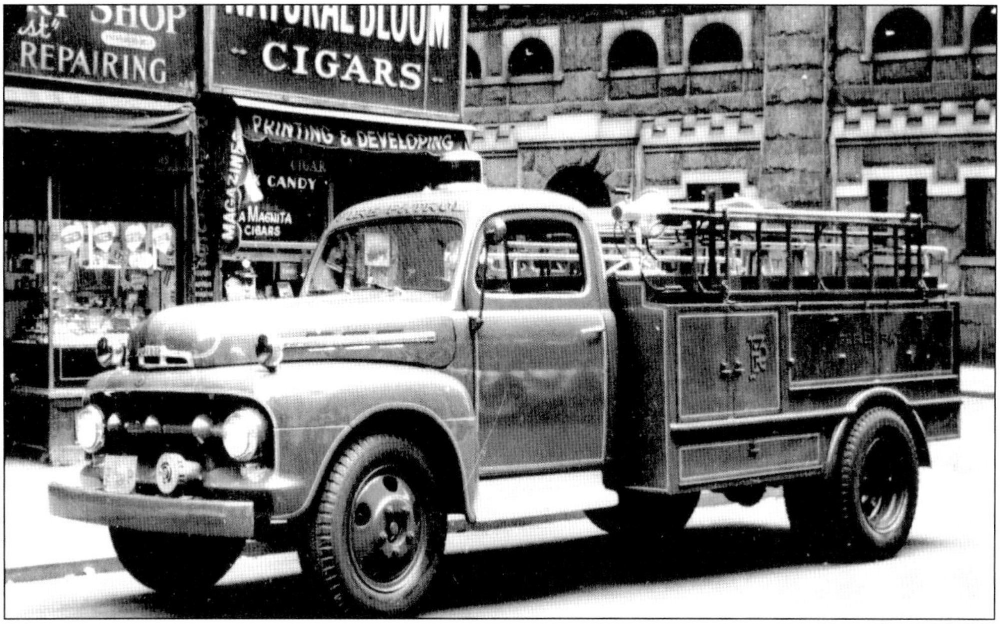

FORDS FROM 1951. During the 1950s these rigs were answering a combined total of over 20,000 alarms in the five boroughs. These trucks were being maintained by the patrol's shop, which moved to Patrol 3 during this time period. By 1960 they were ready for reserve status. (NYFP.)

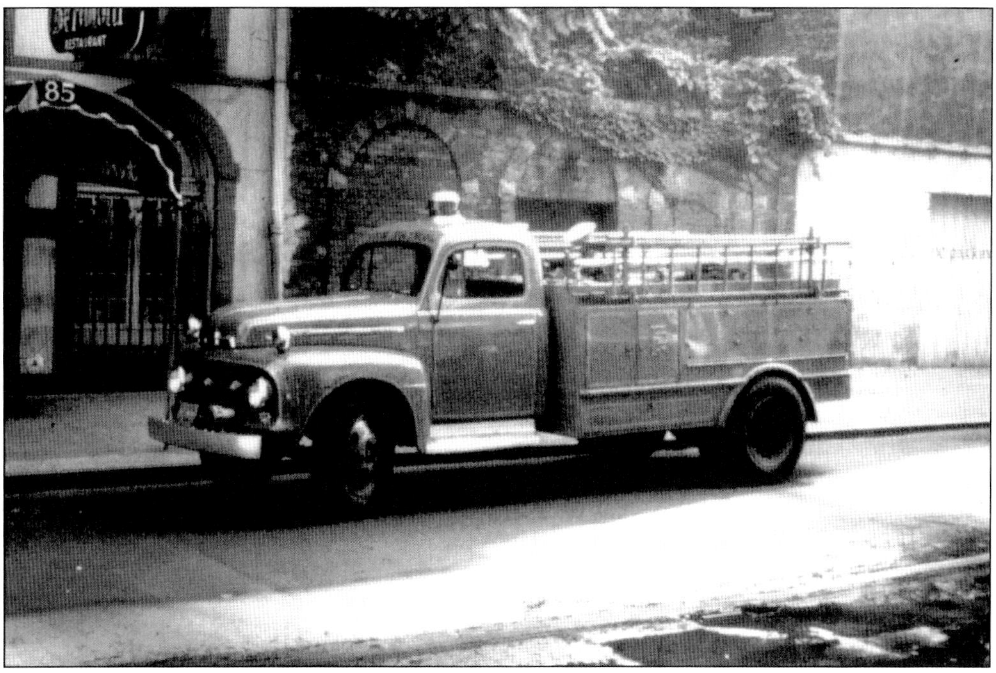

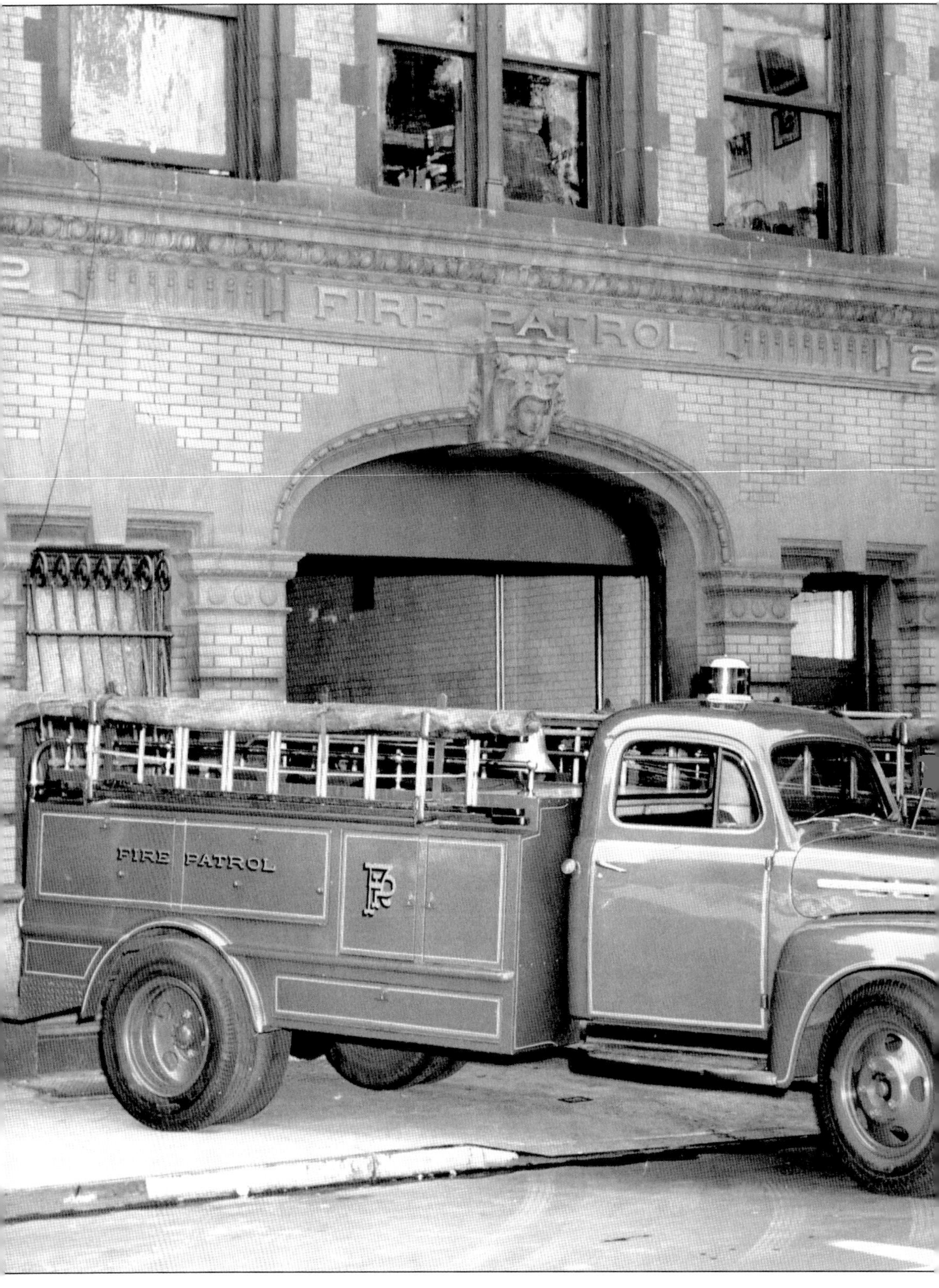
FIRE PATROL
2
2
FIRE PATROL
FP

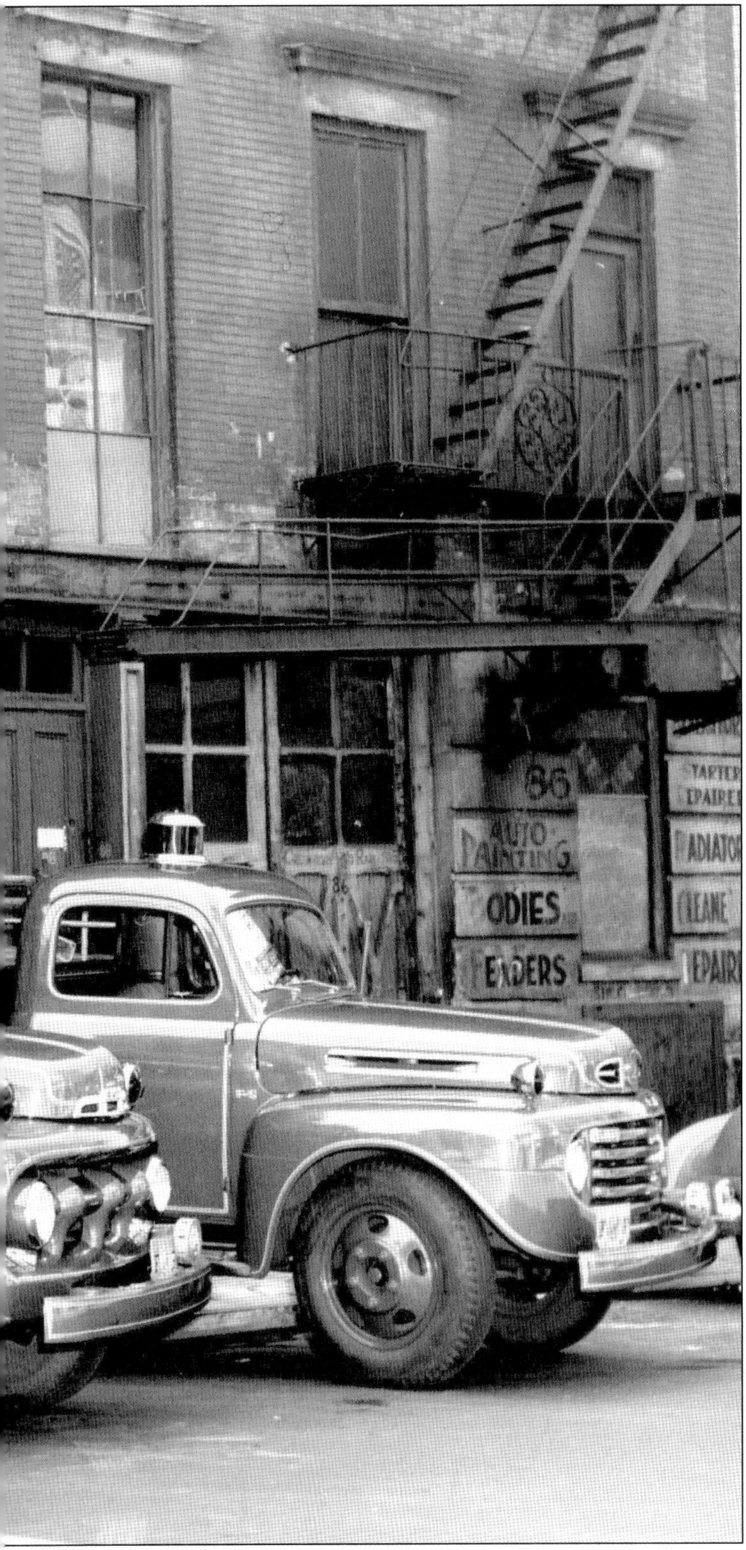

THE FORDS ARRIVE, 1950. In 1950 the patrol began the largest purchase of apparatus since the first round of Whites late in the second decade of the 20th century. Six new Ford F-5 chassis were purchased with six new bodies. At this time, the three 1935 Internationals were sold, four 1934 and 1935 Whites were sold, and one old timer, a 1927 White, which was sold for just $50. Here, a 1949 and a 1951 Ford stand on the apron of Patrol 2.

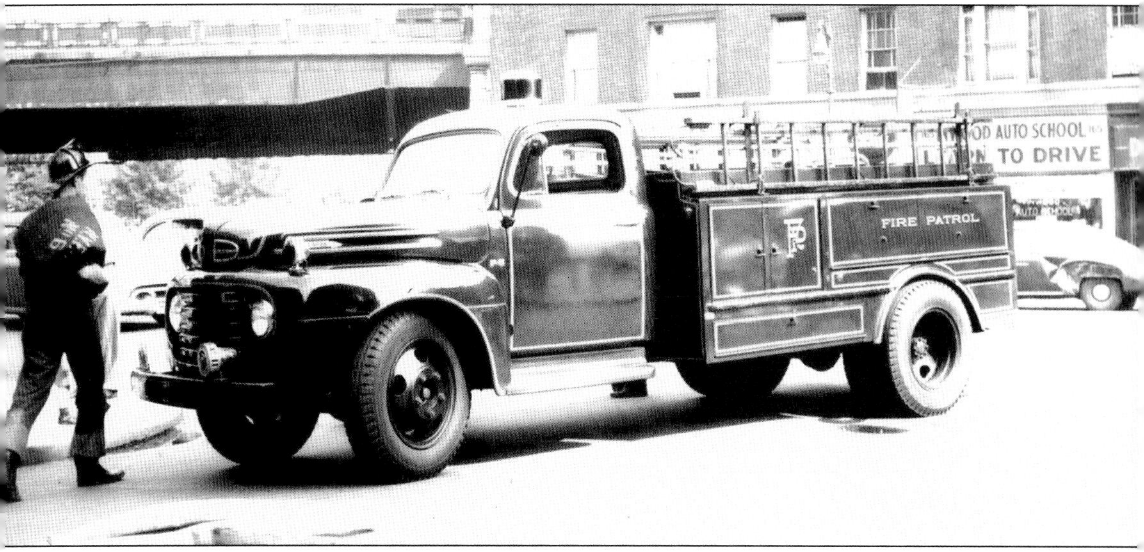

Taking Up at a Bronx Job, 1960. Here a lone patrolman from Patrol 6 is going back to the rig to drop an unused cover off. Standard procedure at all alarms is that each member shoulders two covers along with other tools and equipment. If after the job the covers are not needed, they are returned to the rig. Often unused covers that get wet will be left behind and are known as "sleepers." Wet covers, even if not thrown, are never placed back onto the truck. Wet covers will mold and contaminate the clean covers on the truck. (FDNY.)

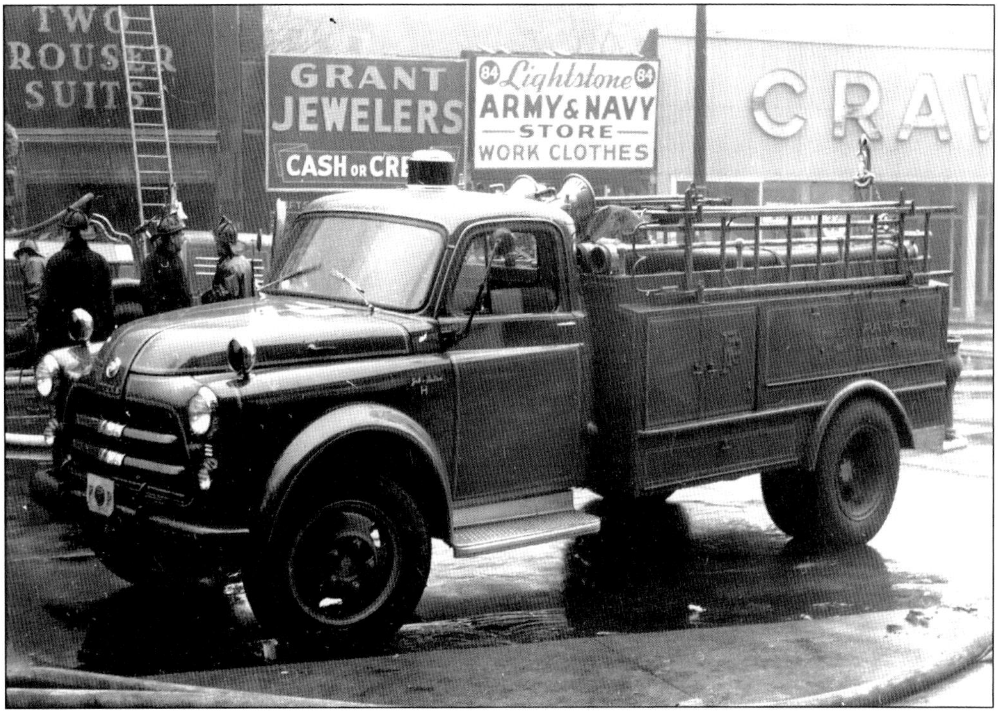

DODGE UTILITY TRUCK. The 1,800-gallon-per-minute pump, which had been purchased many years before, was mounted on this Dodge truck in 1956. Designated as the utility truck, it responded to all flooding conditions and carried special tools that could be available city-wide. In 1964, the utility truck was placed out of service. It would not be until 41 years later that a separate utility truck would be put back in service. (FDNY.)

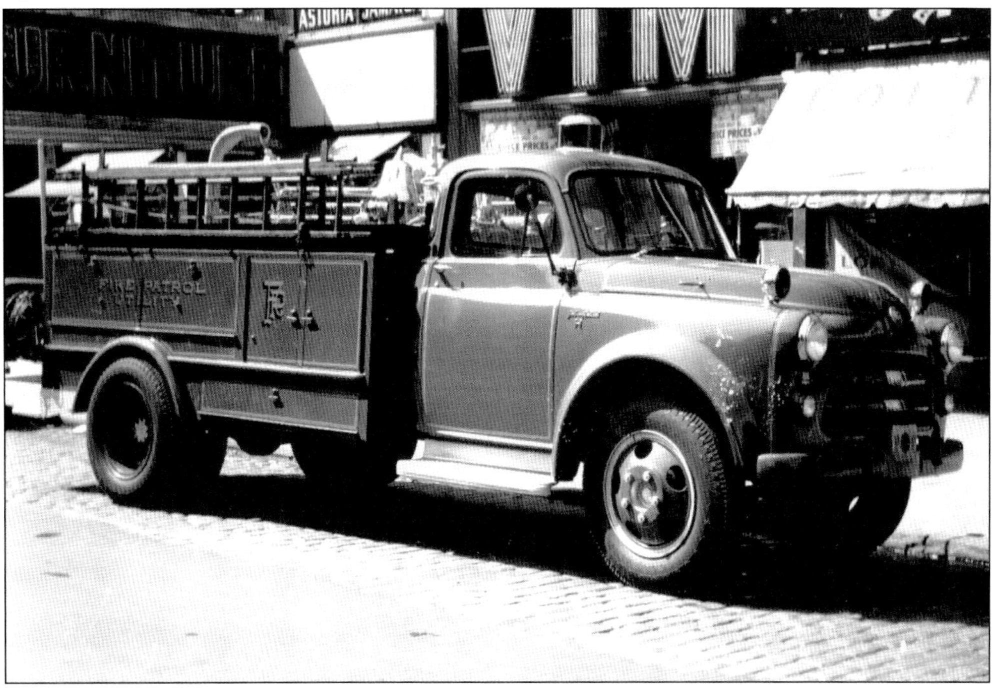

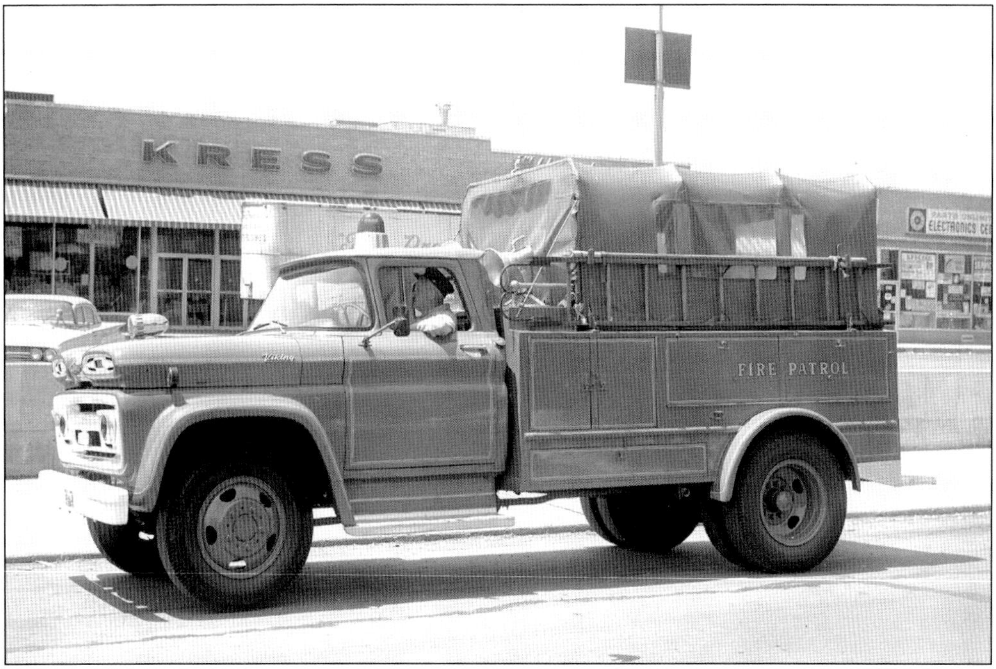

CHEVROLET FROM 1961. Valley Chevrolet delivered five of these 1961 Chevy C-50 series chassis to the patrol. The 1949 bodies were mounted on these trucks, once again extending their lifespan. They were the first rigs with a more modern appearance and most would serve well into the early 1970s as reserve units. (PW.)

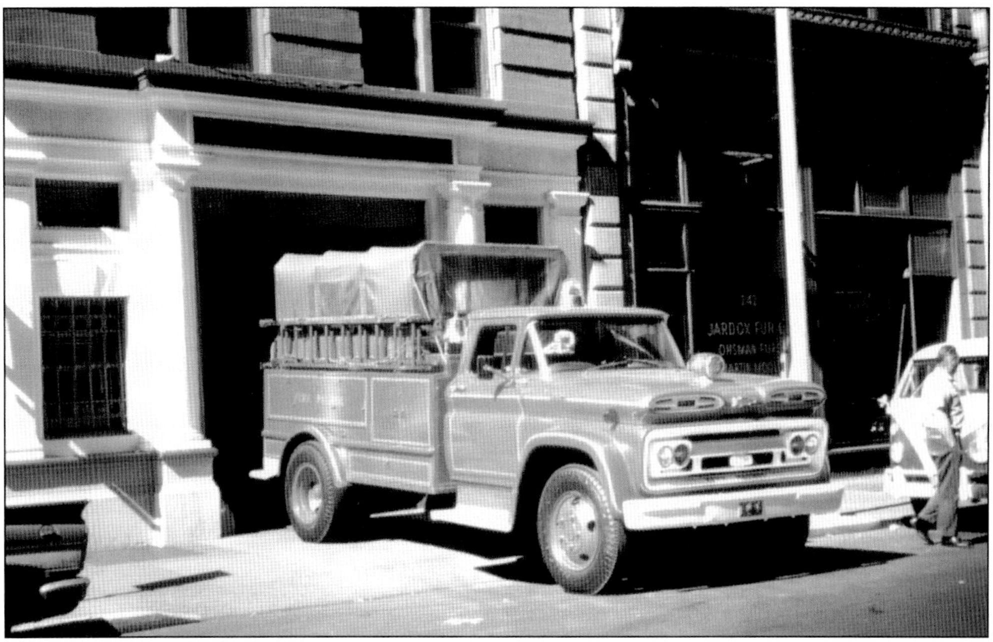

PATROL 1. Fire Patrol 1 received the first of the new Chevy's in 1961. The trucks had Federal Beacon rays mounted on the roof while traffic clearing power consisted of a center mounted Federal siren. The traditional bell was located behind the cab and it is said that it was the duty of the "junior man" to ring the hell out of it while responding. (PW.)

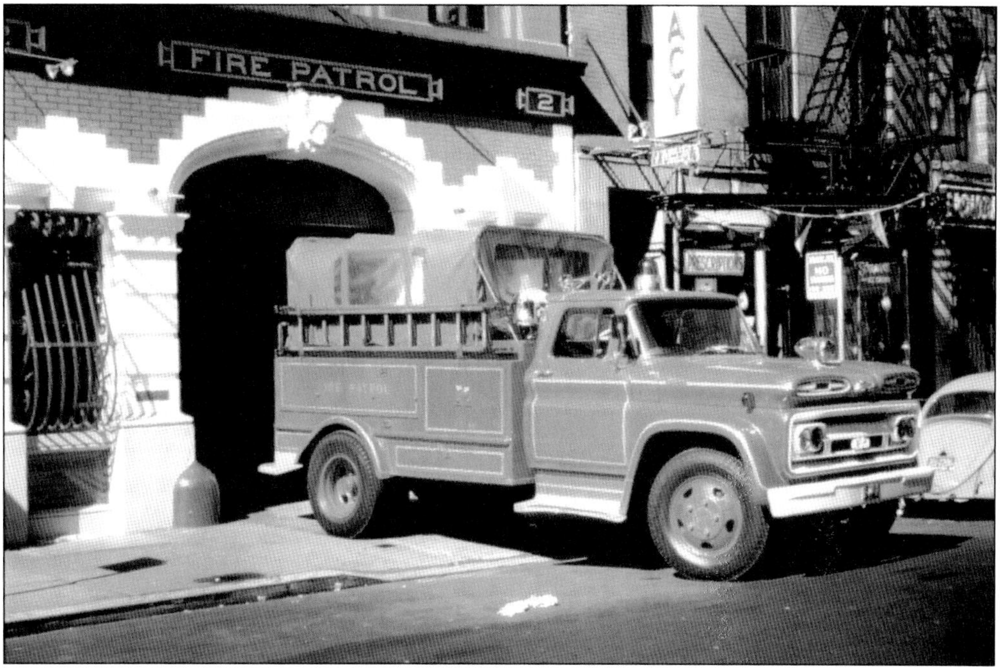

PATROL 2. Next to be delivered to Patrol 2 was a 1961 Chevy. Actually, during this time, Patrol 2 received another 1961 Chevy which was parked behind to be used as a spare. All five rigs were identical except for brass plates which designated the truck number. (PW.)

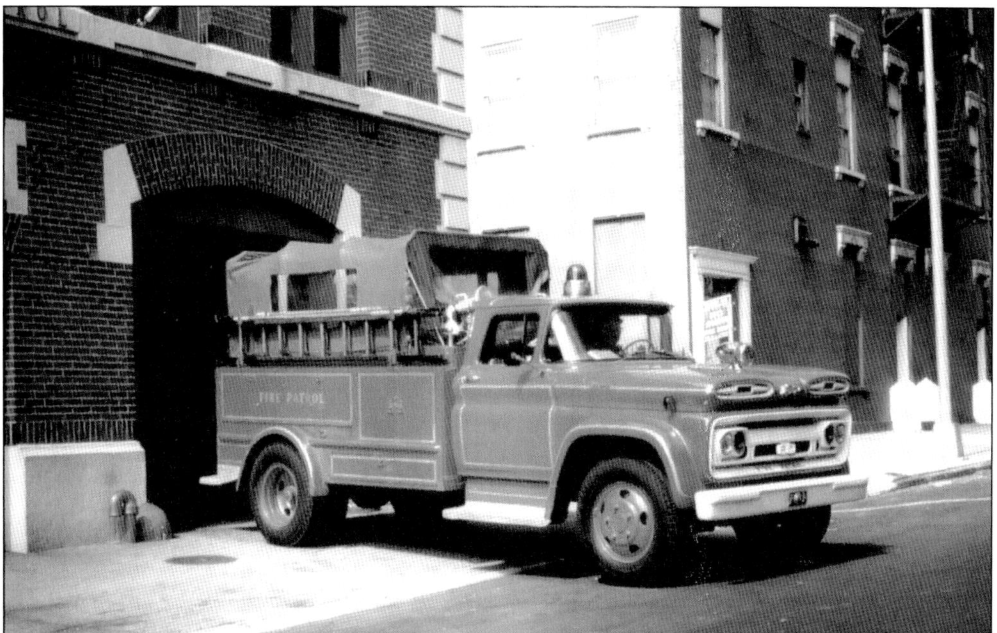

PATROL 3. The 1961 Chevy of Patrol 3 was placed in service at the Dean Street firehouse in the summer of 1961. During that year, the NYFP would respond to more than 7,500 alarms while placing nearly 47,000 miles on the first line apparatus. Designated as Car 48, this rig would serve until 1968. (PW.)

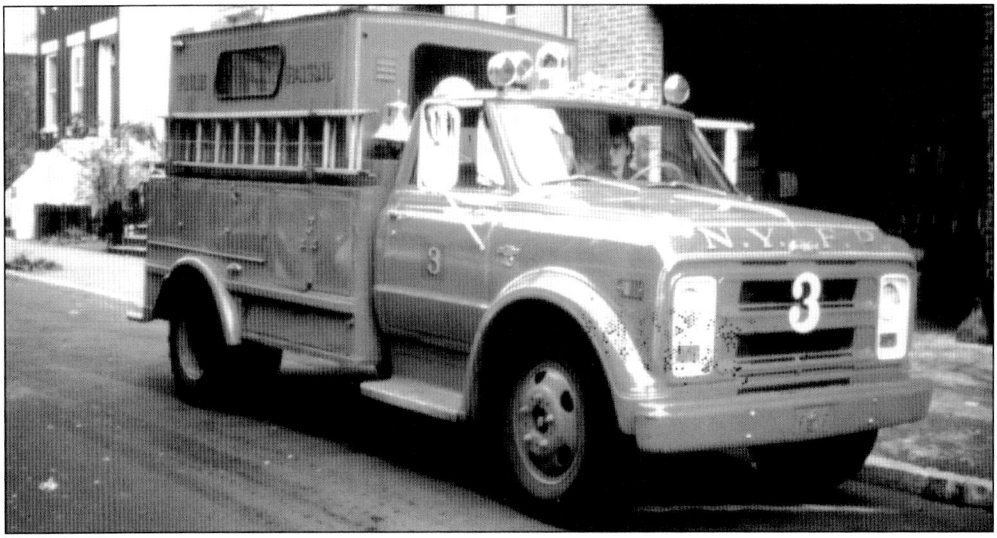

CHEVY PATROL RIG, 1968. In 1968, three of the five 1961 Chevy trucks were placed out of service. Their bodies were placed on the new chassis and plywood enclosures were added for the protection of the men, at this time the rear of the trucks still had no heat. These trucks were the last rigs to be equipped with standard four speed transmissions. (NYFP.)

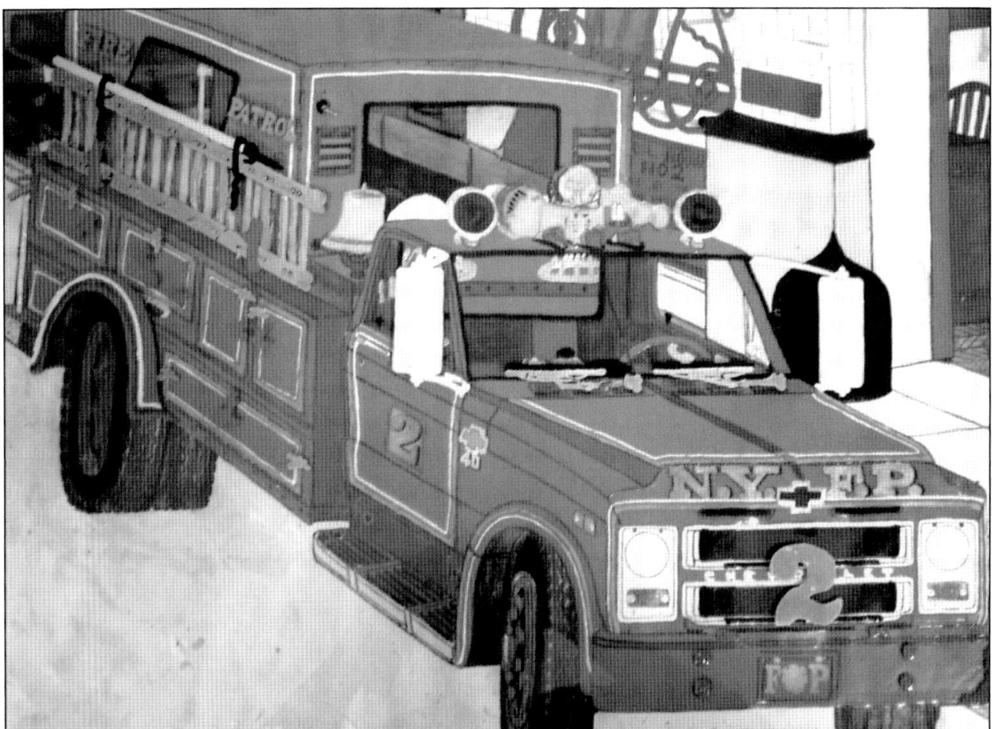

PORTRAIT OF THE RIG. In 1971, Robert J. Martucci painted this three dimensional picture of the 1968 Chevy assigned to Patrol 2. It was presented to his friend, patrolman Paul Curran, as a token of his appreciation. Shortly after, the truck was placed out of service and patrolman Curran left for New York City Transit. Upon his retirement from the MTA, Curran returned to the fire patrol and at the time of this publication is still responding to fires at Patrol 3. (PC.)

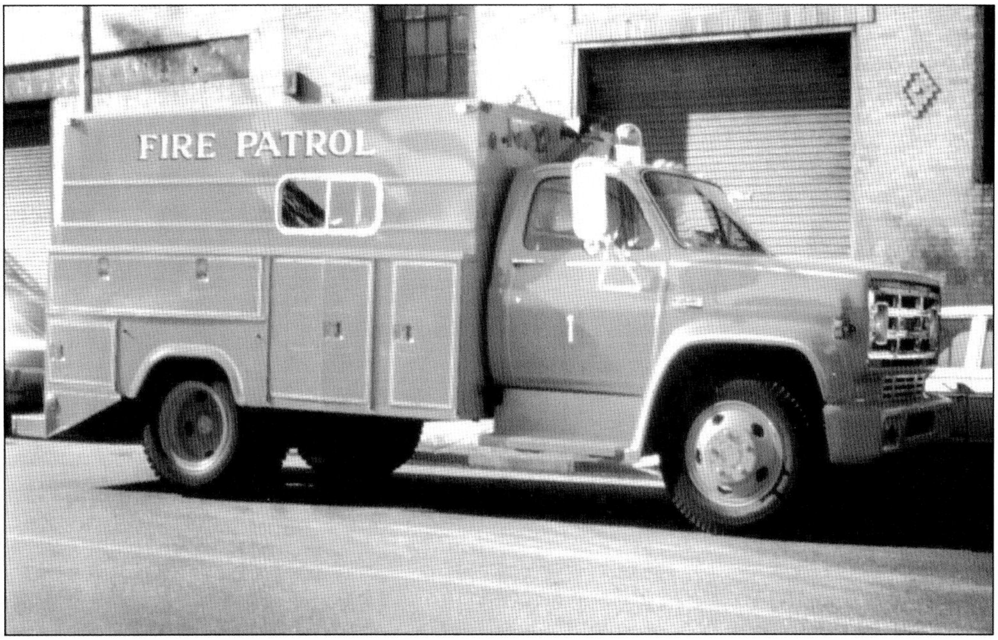

A New Design for the 1970s. Starting in 1973, the first of a long line of GMC trucks were purchased. After many years of crudely attempting to enclose the bodies, these new trucks were delivered with completely enclosed and heated rescue style bodies. The bodies, which were built by the Independent Truck Company, contained eight large exterior compartments plus interior compartments and bench seats. During this time, the patrol was riding with an average of seven men per tour. (PW.)

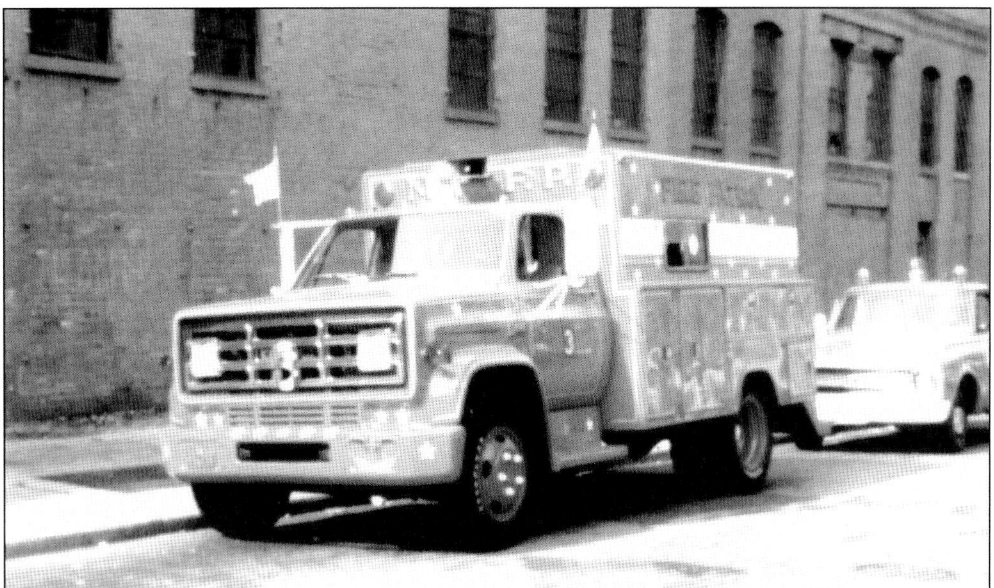

Bicentennial Paint. Starting in 1975, nearly every piece of fire apparatus in New York City was decorated in some type of red, white, and blue paint scheme. Patrols 1 and 2 each painted stars and stripes all over the rig while Patrol 3 painted the entire box of their truck in a flag layout. Even the houses were painted in ornate patriotic colors. (PW.)

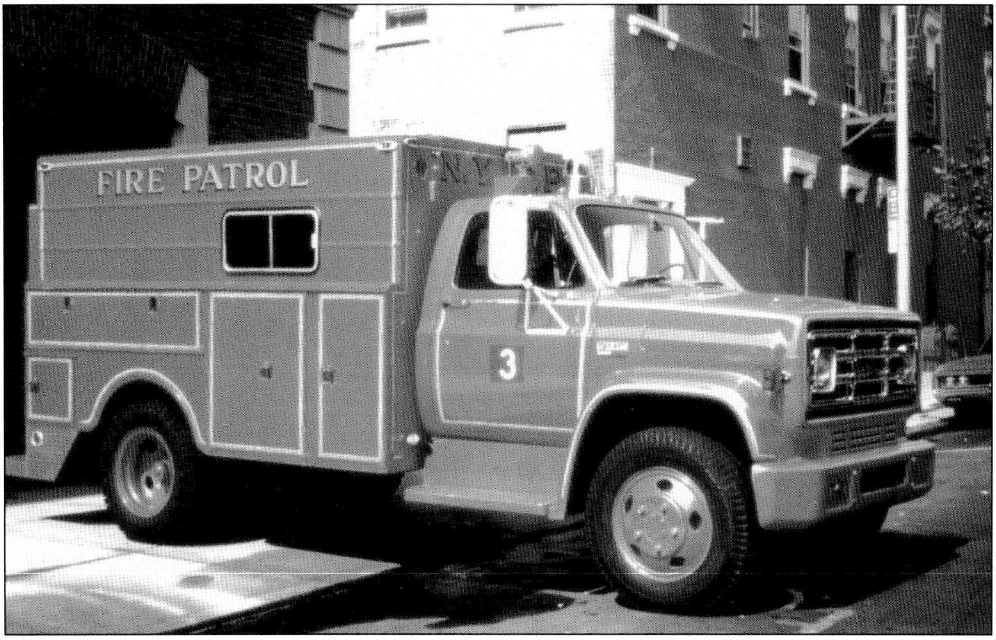

GMC, 1973. Each of the new trucks had large gilt lettering on each side as well as elaborate gold pin striping on the cab and body. Each truck was equipped with a roof mounted Federal Beacon ray, and a Federal model 28 siren in order to get city traffic to yield. (PW.)

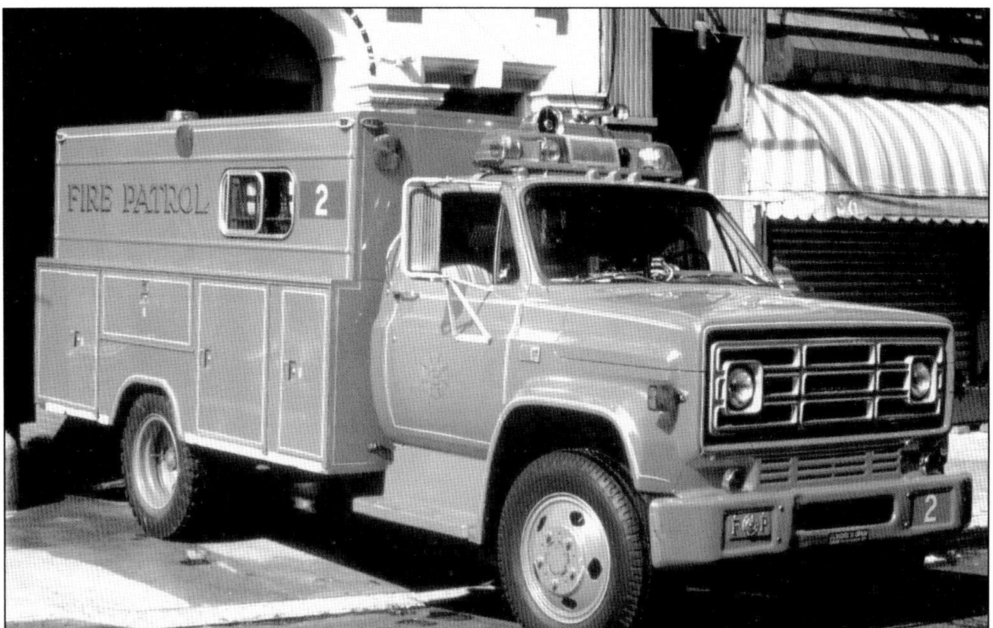

GMC PATROL 2, 1981. In early 1981, Patrol 2 placed their 1973 GMC into reserve status. Although similar in appearance, this 1981 model had a rear box of different design. The new bodies, which were built by Stahl, were about eight inches lower than before and the compartments were of a slightly different configuration. Each of these rigs could carry 58 covers, five roof covers, two cellar pumps, and other assorted salvage equipment. (JK.)

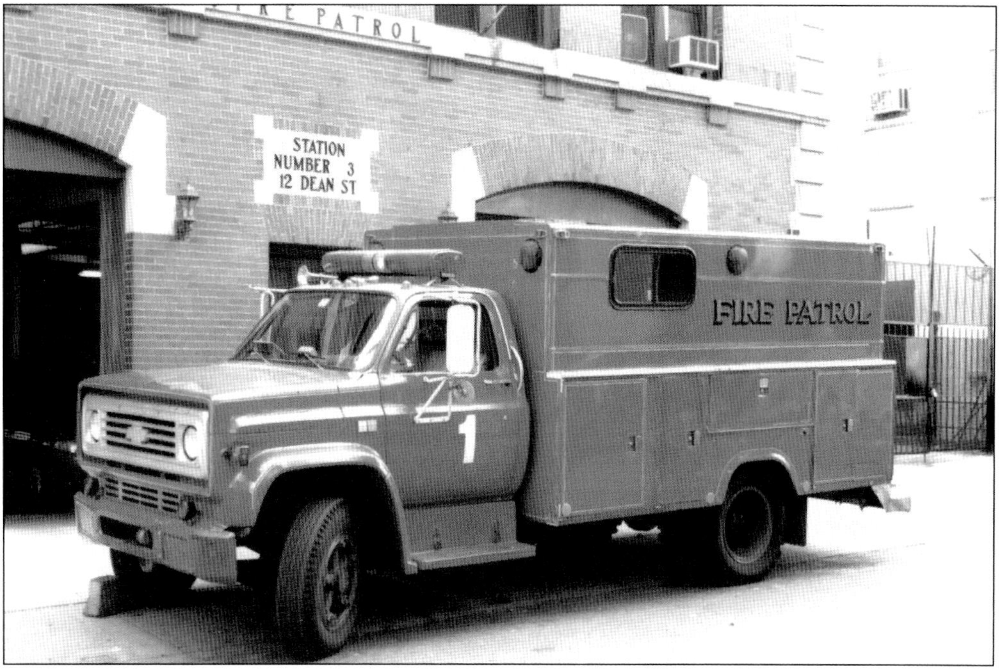

CHEVROLET, 1986. Patrol 1 took delivery of this rig to replace their earlier vintage GMC. This new rig would serve the busiest patrol company in the United States. It would serve for many more years, eventually becoming the spare rig for all three companies. (JK.)

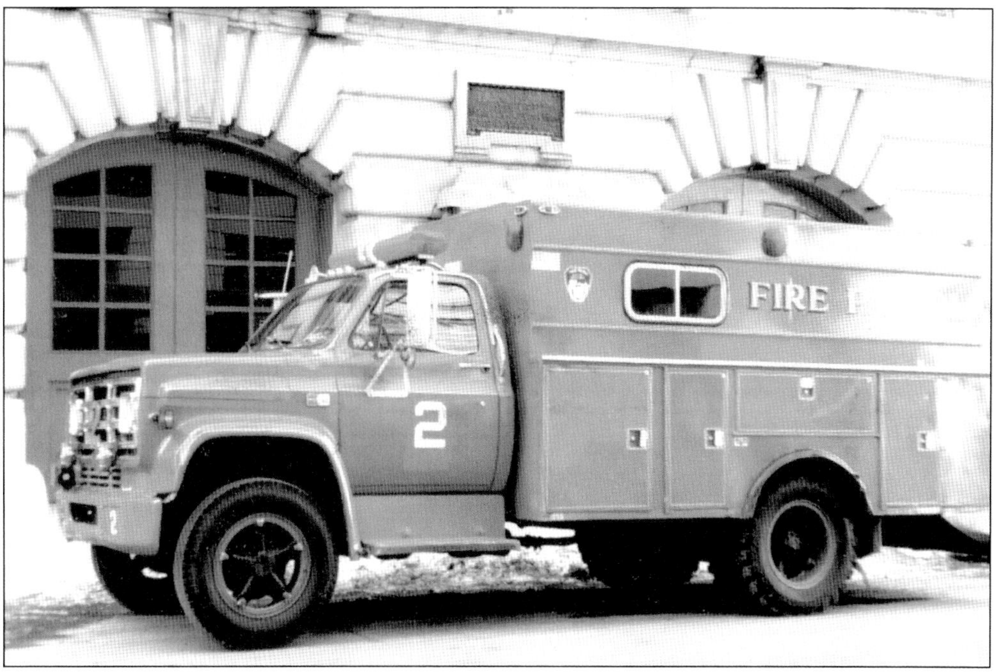

GMC PATROL 2, 1987. In 1987, the last group of conventional GMC 7500 series trucks was delivered. These new trucks were all equipped with Federal Aero Dynic lightbars, side body flashing lights, and air horns. In keeping with tradition they were all painted Fire Patrol red. (JK.)

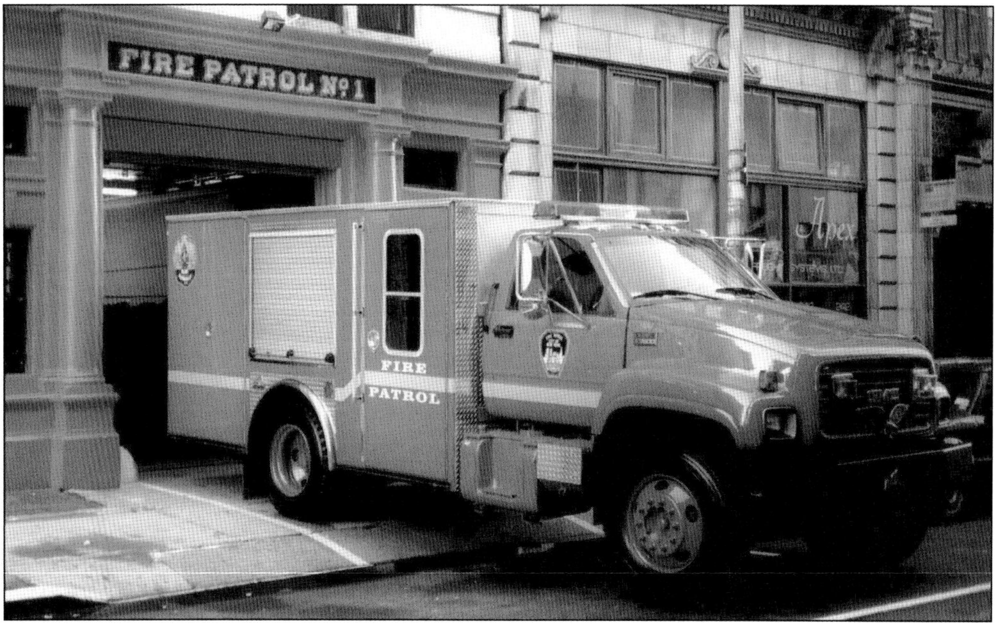

NEW RIGS FOR A NEW CENTURY. In 1999, the fire patrol, along with the help of Ferrara Fire Apparatus in Holden, Louisiana, began radical specifications for three new rigs. The new rigs became the first diesel powered units in the history of the patrol and also featured longitudal self contained breathing apparatus seats for a crew of four men. Here, Car 21 exits quarters at their historic firehouse on West 13th Street in the shadow of the Empire State Building. (JK.)

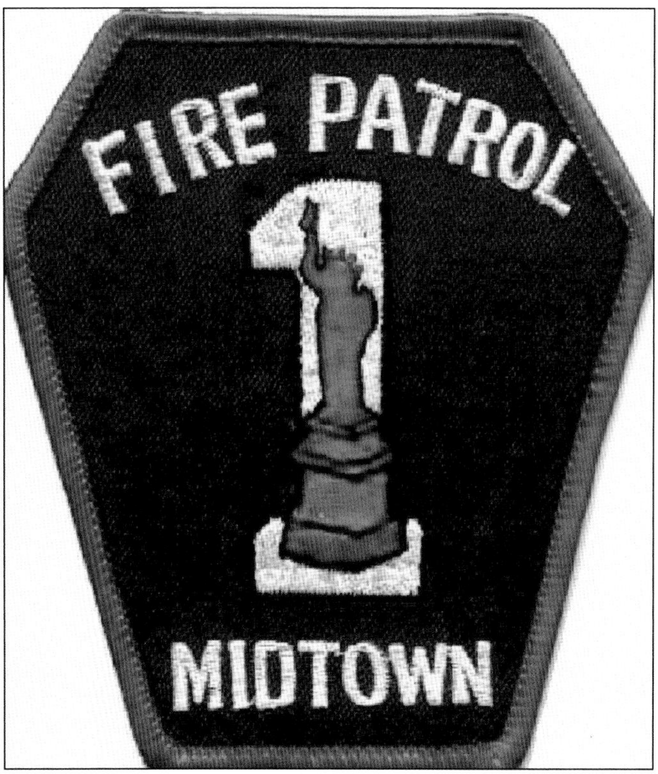

COMPANY PATCH OF PATROL NO. 1, 1970S. The first company patches were designed in the mid-1970s. Fire Patrol 1 took this design with the Statue of Liberty ensconced in a large numeral one. This patch was later replaced by a Maltese cross with a large blue person, resembling the Campbell's Soup Kid. The motto of Patrol 1 is "We got ya covered."

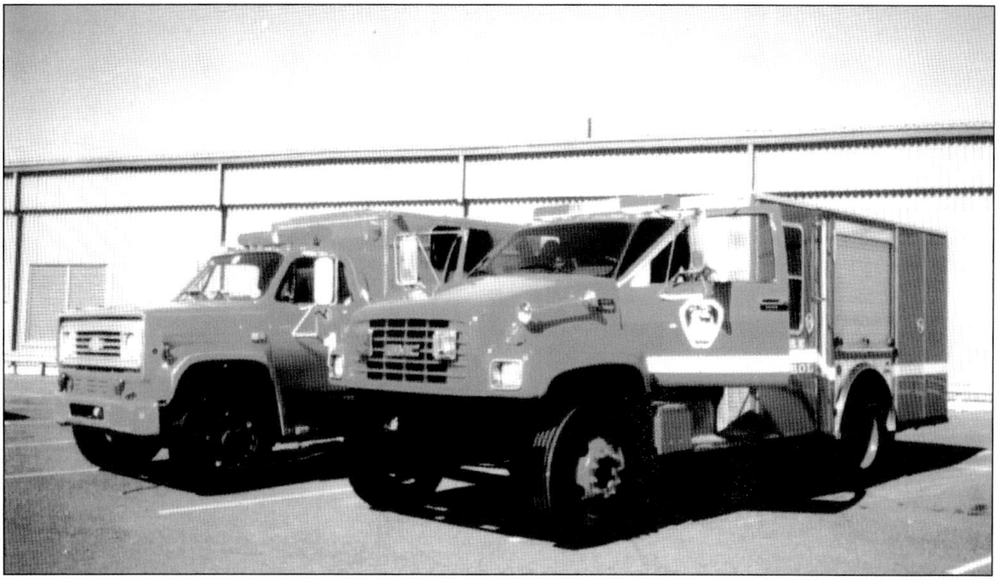

The Old and the New. In late 1999, the first of the new GMC/Ferrara trucks were placed in service. Here, the new truck of Patrol 2 is photographed with their Chevy/Stahl patrol wagon along the Brooklyn waterfront. The 1999 truck was seriously damaged on September 11, 2001. (JK.)

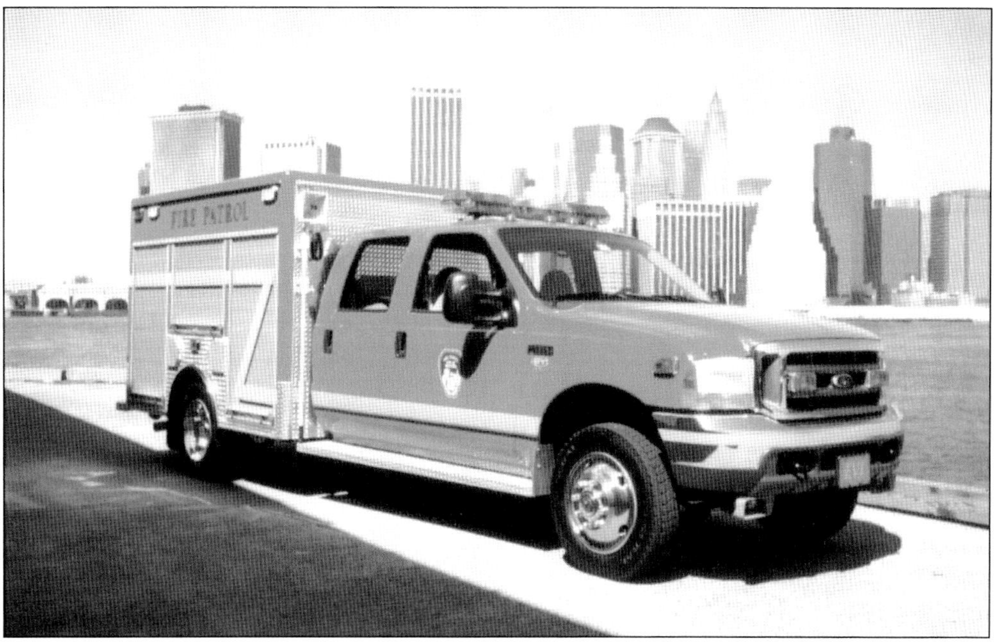

Car 52. Shortly after September 11, 2001, the NYFP took delivery of this Ford F-550 with bodywork by EVI located in Florida. Powered by a diesel engine, it contains an onboard generator, quartz scene lighting, and seating for five men. In 2005, this truck was redesignated as the utility truck, the first time the NYFP has operated one since 1964. (JK.)

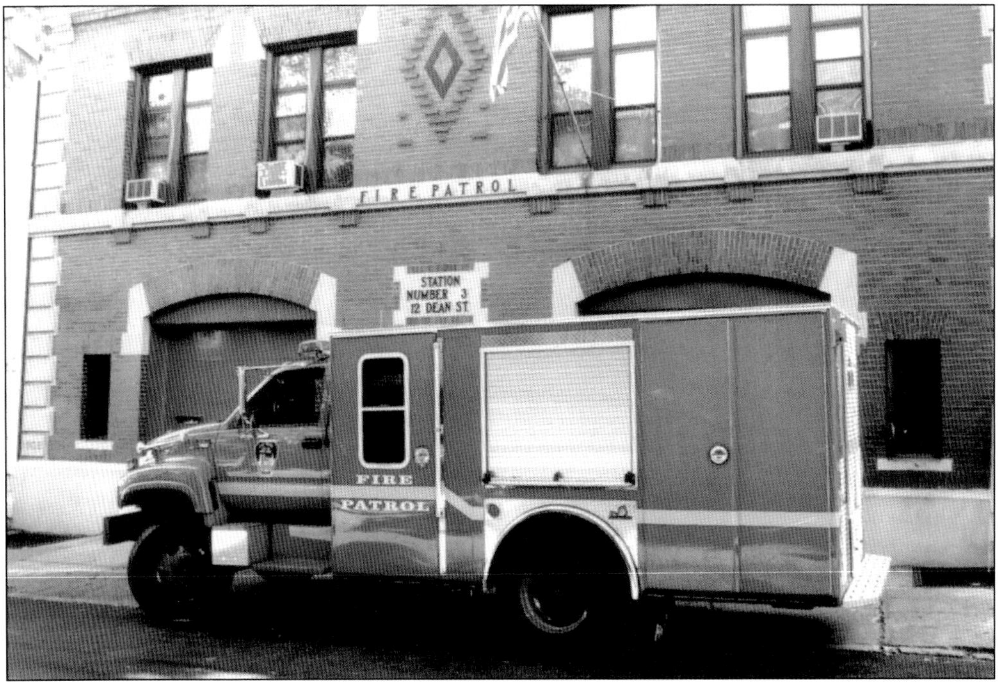

PATROL 3. All of the 1999 GMC 7500 series trucks were identical. They were originally delivered without company numbers, but in 2001 the numbers were applied on the crew compartment doors. Company patches are also part of the graphics package. In all, each of these trucks carry 48 stock covers, four roof covers, assorted sheet plastic, lightweight vinyl covers, a generator, four cellar pumps, a large 450-gallon-per-minute pump, forcible entry equipment, and six SCBAs. (JK.)

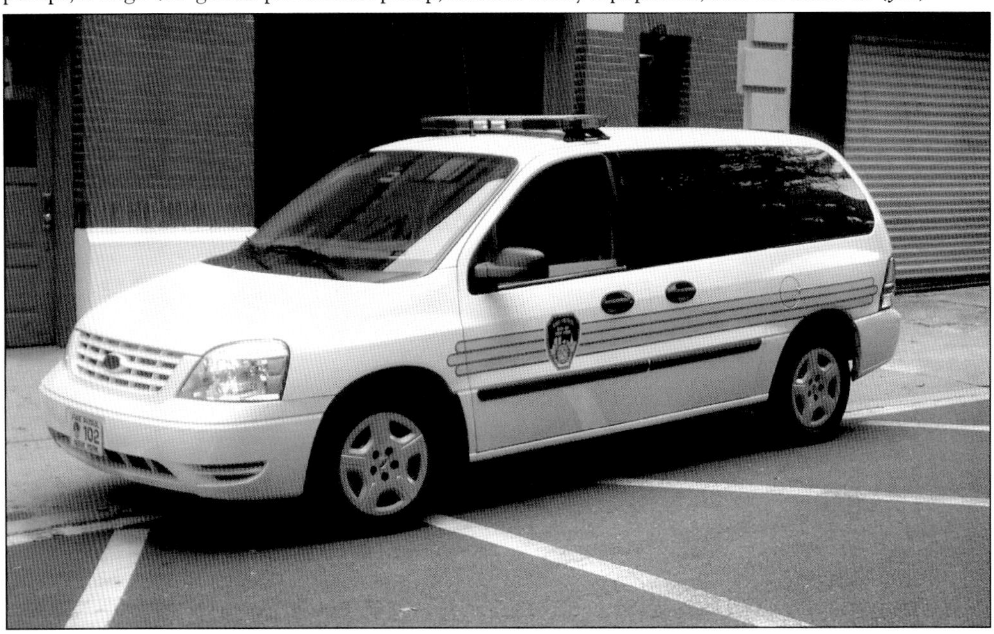

CAR 102. After the former fire prevention vehicle was destroyed in an accident, this minivan was placed in service in 2005. It is the first white painted vehicle of the patrol and is used by both the fire prevention division and the training division. (TER.)

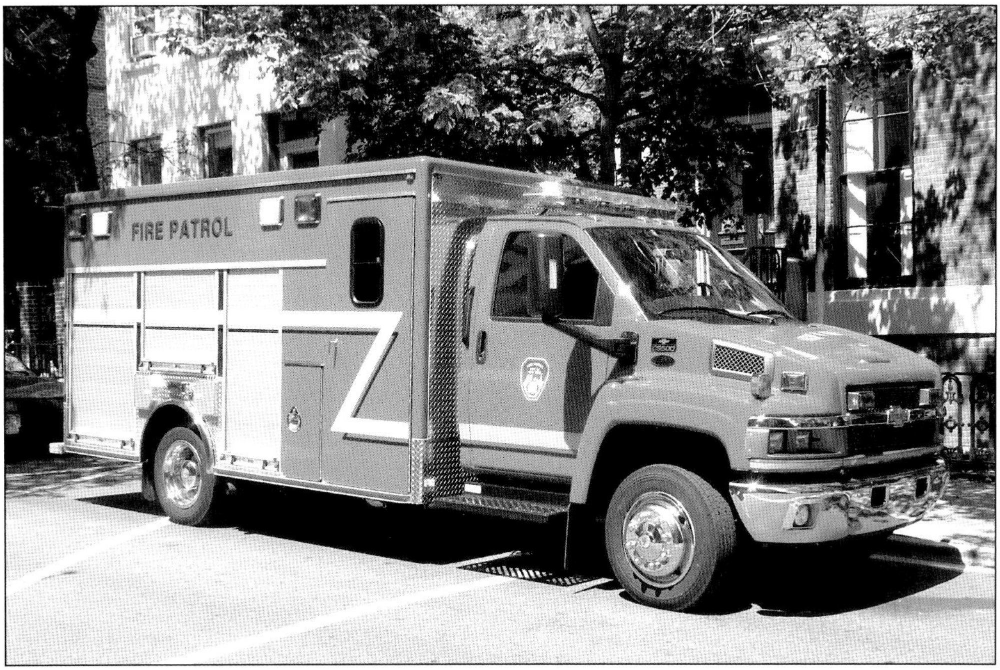

FIRE PATROL 3. As this book went to publication, the NYFP took delivery of the largest apparatus ever to be placed in service by the department. It is a 2005 Chevrolet/EVI salvage unit with a 16-foot body. Extensive scene lighting, roll up doors, and front and rear air conditioning round out this vehicles features.

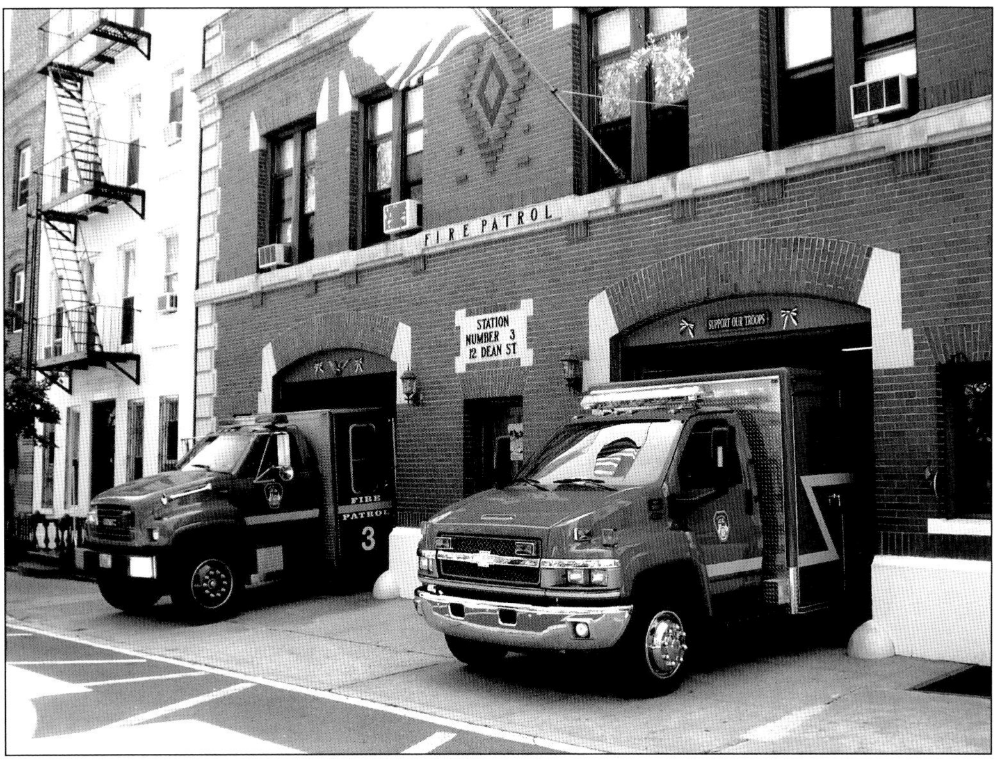

BIBLIOGRAPHY

Burness, Tad. *American Truck and Bus Spotters Guide*. Osceola, Wisconsin: Motorbooks, Inc., 1985.
Costello, A. E. *Our Fireman*. New York, New York: Knickerbocker Press, 1887.
Cull, Frank. *The 23rd Street Fire As it Happened*. New York, New York, 1991.
Douglas, Emmons. *While the Flames Raged*. Middletown, New York: The Engine House, 1994.
Dunshee, Kenneth. *As You Pass By*. New York, New York: Hastings House, 1952.
Gispert, Simon. *Salvage Corps and Fire Patrols*. London, England, 1999.
Homberger, Eric. *The Historical Atlas of New York City*. New York, New York: Owl Book, Henry Holt, 1994.
King, William T. *History of the American Steam Fire-Engine*. Mineola, New York: Dover Pub, 1896.
Smith, Arthur. *The New York City Fire Patrol*. Reading, Pennsylvania: Exeter House, 1997.
WNYF Magazine. New York, New York: Fire Department New York, 1956–1992.
New York Times Archives. Articles. New York, New York, 1853-2005.
New York Board of Fire Underwriters Minutes. New York, New York: unpublished, 1851–1870.

Photograph Credits
(PC) Paul Curran
(PW) Paul Wormsely
(LW) Larry Wagonfeld
(JK) Joe Krasko
(DK) Dennis Kenney
(MK) Mike Krempasky
(TM) Tom McGee
(LM) Larry McGuiness
(BM) Brian Mulvey
(SM) Simon Gispert
(TER) Tim Regan
(EM) Ed Moran
(JL) Jack Lerch
(FL) Frank Lennon
(JS) John Sheehan
(RH) Rich Heffernan
(AT) Al Tronjaniwcz
(NYFP) New York Fire Patrol
(FDNY) Fire Department New York
(UFPA) Uniformed Fire Patrolman's Association